Collecting, Processing and Germinating Seeds of Wildland Plants

COLLECTING, PROCESSING AND GERMINATING

Seeds of Wildland Plants

James A. Young & Cheryl G. Young

TIMBER PRESS
Portland, Oregon

We wish to express our appreciation to the scientists and technicians who contributed to the various editions of *Collecting, Processing and Germinating Seeds of Wildland Plants*. Special thanks to Howard Sherman and Jean Stoess for editorial assistance and encouragement. We appreciate the effort extended by Margaret Williams of the Northern Nevada Native Plant Society in reviewing the manuscript.

TIMBER PRESS
9999 SW Wilshire
Portland, Oregon 97225

Contents

CHAPTER 1

Introduction

The growing of native plants to enhance the environment is a very worthwhile objective. The environment where you want to culture the native plants may be the garden surrounding your home or thousands of acres of wildlife habitat.

Many gardeners are introduced to the growing of wildland plants on a spur-of-the-moment basis. While hiking, fishing or otherwise enjoying the outdoors you happen to see an attractive wildland plant. If the plant happens to be in seed, you may pick a few seeds and deposit them in your pocket or a sandwich bag. Months may pass before the forgotten seeds are rediscovered and the thought of germinating them is reawakened.

We do not mean to downplay the joy of chance encounters with interesting new plants. However, your chances of successfully propagating such plants can be greatly enhanced by a little forethought and preparation. This volume is designed to provide guidelines so you can be prepared to collect seeds of interesting plants. These collections need to be properly timed to insure optimum viability. Seeds or fruits have to be threshed and cleaned before they can be germinated. Often it is necessary to store the seeds for a considerable period of time. Most importantly of all, the seeds of many wildland species are dormant and require pretreatment to obtain germination.

We have updated and consolidated in one presentation the standard publications on seeds of wildland species that have been

1

available to gardeners. The updating is based on a review of the recently published scientific literature and the extensive research efforts of the authors and their colleagues.

Gardeners may be shocked to find we have included information on the germination ecology of numerous species of weeds. Knowledge of the ecology of weed species is the first and most necessary step in developing truly integrated pest management procedures. For many genera or even some families of wildland plants the only information we have for germination ecology of the seeds comes from study of weedy species.

All of this may sound complicated, but with a little effort the growing of wildland plants from seeds is relatively easy.

CHAPTER 2

What is a Seed?

Seeds are such commonplace items it is easy to forget what a seed represents. All of the higher or more fully evolved plants produce seeds. These plants are known as angiosperms, or seed-bearing plants. Seeds consist of an embryo or miniature plant, usually some form of stored food to nourish the embryo, and a protective covering.

If you doubt the importance of seeds, look closely at your weekly market basket. Bread, beans, rice, nuts, alfalfa sprouts, mustard, the list of seed-derived foods is limited only by the time one is willing to spend adding to the total. Although seeds are basic to our food supply, we are interested in this volume in using seeds to create new plants.

DEVELOPMENT OF SEEDS

Seed development starts with the flowering process. A pollen grain alights on the stigma or receptive surface of the female flower parts, and a pollen tube containing two sperm cells begins growth down the style to the ovary. Seeds develop from the fertilization or joining of one of the sperm cells with the female egg in the ovary. A seed consists of an embryo, or miniature plant, surrounded by an envelope of tissue. The embryo is derived or is the result of the fertilization of the egg cell in the embryo sack by the male sperm cell from the pollen tube. The envelope of tissue or testa that surrounds the embryo originates from the mother plant from tissue surrounding the ovule called integuments. The testa eventually will become the seed coat, which is so important in the germination ecology of seeds.

Endosperm

Many seeds contain in addition to an embryo and seed coat an endosperm. The endosperm is derived from both parents in the case of cross-pollinated plants, and from the same plant in self-pollinated plants. The endosperm is formed as a result of the triple fusion of the polar nuclei with the second sperm nucleus from the pollen tube. The polar nuclei are the egg cells that existed in the ovary that were not fertilized by the other sperm cell from the pollen tube.

There is great variability among plant species in the development and importance of the endosperm. The endosperm is not always present. In many plants, it is greatly reduced or as in the case of orchids the formation of the endosperm is completely suppressed. In seeds where the endosperm develops it frequently contains material utilized by the embryo during germination as a source of energy. The endosperm may contain a variety of storage material such as starch, oils, proteins, or hemicellulose. Many of the seeds that form the vital foodstuffs for man, such as rice, wheat, and corn, store energy for germination in their endosperms.

Embryo

The embryo in a seed consists of a radicle or root, a plumule or shoot, one or more cotyledons or seed leaves, and a hypocotyl that connects the root and shoot. The flowering plants or angiosperms are divided into two great classes based on the number of cotyledons: the Dicotyledoneae with two cotyledons, and the Monocotyledoneae with one.

The cotyledons or seed leaves may be a source of food reserve for the developing embryo, or serve the dual purpose of food storage and act as photosynthetic surfaces to produce food for the seedling. Seedlings are classified as epigeal, in which the cotyledons are above ground, and hypogeal, in which the cotyledons remain below ground during germination. When the cotyledons are epigeal they are pulled through the soil surface by the epicotyl arch.

The embryo may be variously located within the seed and may either fill the seed almost completely as in the rose and mustard families, or it may be almost rudimentary as in the buttercup family. The classification of seeds is often based on the size and position of the embryo and on the ratio of the size of the embryo to that of the storage tissue.

In many plants special absorbing tissues develop which take up the reserve materials from the endosperm. In the onion the tip of the cotyledon is actually embedded in the endosperm and withdraws the food reserves from the endosperm. Onions are epigeal in germination with the single cotyledon visible above the seed bed surface. The onion cotyledon becomes green when exposed above the soil surface.

In the grass family specialization has proceeded further, and a special root-like structure called a scutellum withdraws substances from the endosperm to provide food for the embryo during germination. An example of this type of hypogeal germination is corn or wheat.

Such complications are absent in most of the dicotyledonous plants. The seedlings of castor bean are an example of epigeal germination with an endosperm. The common French bean seedling results from epigeal germination without endospermic nutrition. The French bean seedlings have the characteristic epicotyl arch which pulls the cotyledons through the soil surface.

Testa and Fruits

The testa or seed coat varies greatly in form. It may be soft, gelatinous or hairy, although a hard seed coat is the form most commonly met. Seeds are formed in the ovary, and the mature ovary is a fruit. Fruits arising primarily from the ovary are known as true fruits, while fruits in which other structures or several ovaries and their related structures participate are often termed false fruits. Both types of fruits may be dry or fleshy.

In many plants the testa and the ovary wall are completely fused so that the seed and fruit are in fact one entity, as is the case in the grains of the grasses. The fruit of a grass which we commonly refer to as a seed is actually a caryopsis. The small, dry, hard, and indehiscent fruits of many dicotyledonous plants that contain a single seed are called akenes.

Quite simply, a seed consists of an embryonic plant composed of a miniature root, shoot, and connecting tissue. This plant is enclosed within a protective cover composed of the seed coat and often includes all or portions of a surrounding fruit. Within the seed coat along with the embryonic plant is often found some form of food storage to provide energy for germination. The food is stored either as endosperm or in cotyledons. In handling seeds during collection, threshing, storage, or germination it is vital to remember that within the seed coat there is a living plant.

SEED GERMINATION

Later we will devote an entire section to germination and dormancy. However, before we start collecting and processing seeds, it is desirable to review what triggers germination and what processes are involved in germination.

Seed germination is the resumption of active growth of the embryo that results in the rupture of the seed coat and the emergence of the young plant. This definition presumes that the seed has been in a state of rest after its formation and development. During this period of rest, the seed is in a relatively inactive state and has a low metabolic rate. The seed can remain in this state until the time and place are right for resumption of growth.

Some seeds are capable of germination soon after fertilization and long before their normal harvest time, while others may be dormant and require an extended rest period or additional development before germination can occur. Seed germination can be characterized by several processes.

Processes in Seed Germination

The major events occurring in seed germination are: water imbibition, enzyme activation, initiation of embryo growth, rupture of the seed coat, and emergence of the seedling. Germination, if successful, is followed by seedling establishment.

Water imbibition. Water is absorbed through natural openings in the seed coat, and diffuses through the seed tissues. The imbibition of water causes the shrunken cells in the relatively dry seed to become swollen with liquid or, in botanical terms, to become turgid. The entire seed grows in volume; the seed coat becomes more permeable to oxygen and carbon dioxide. As swelling occurs, the seed coat often ruptures, facilitating both water and gas uptake, and the emergence of the growing points.

Enzyme activation. Water absorbed in seed tissue activates the various enzyme systems which serve to: (1) break down stored tissue, (2) aid in the transfer of nutrients from storage areas in the cotyledons or endosperm to the growing points, and (3) trigger chemical reactions which use breakdown products in the synthesis of new material.

Initiation of embryo growth. Following enzyme activation, new material begins to be synthesized, reflected by an increase in the size of the embryo.

Seedling emergence. The primary root is usually the first structure to emerge from the seed coat. This is ecologically very advantageous because it gives the seedling a chance to establish early root connection with the moist soil.

For many grass seeds the first structure to emerge is a covering of the radicle, termed coleorhiza. The amount of coleorhiza varies greatly among species. The function of this tissue may be to aid in the transfer of moisture from the seed bed to the seed.

SUMMARY

Although seeds vary in size and shape, they all contain a miniature plant in the form of an embryo. Besides the embryo, seeds often contain food stored as endosperm or cotyledons. The embryo and stored food are protected by a seed coat. Seeds are relatively dry. The germination process, in the absence of dormancy, is triggered by the availability of moisture in the seedbed.

SUGGESTED ADDITIONAL READING

Bhztnager, S. B. and B. M. Johri. 1972. Development of angiosperm seeds. Pp. 77–149. *In Seed Biology* Vol. I. (ed.) T. T. Kozlowski. Academic Press, New York and London.

Brink, R. A. and D. C. Cooper. 1947. The endosperm in seed development. *Botanical Review* 13:423–541.

Brown, B. V. 1960. The morphology of the grass embryo. *Phytomorphology* 10:215–223.

Chute, H. B. 1932. The morphology and anatomy of the achene. *American Journal of Botany.* 17:703–723.

Copeland, L. E. 1976. *Principles of seed science and technology.* Burgess Publ. Co., Minneapolis, MN. 369 pp.

Leininger, L. N. and A. L. Urie. 1964. Development of safflower seed from flowering to maturity. *Crop Science* 4:83–87.

Mayer, A. M. and A. Poljakoff-Mayber. 1963. *The germination of seeds.* Pergamon Press, Oxford. 236 pp.

Singh, B. 1953. Studies on the structure and development of seeds of *curcurbitaceae. Phytomorphology* 3:224–239.

Randolph, L. F. 1936. Developmental morphology of the caryopsis in maize. *Journal of Agricultural Research* 53:881–916.

Varner, J. E. 1965. Seed development and germination. Pp. 763–792. In *Plant Biochemistry* (ed.) James Benner and J. E. Varner. Academlc Press, New York.

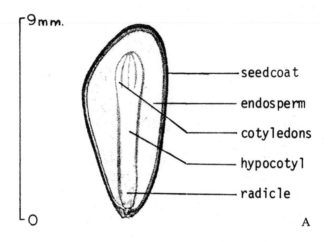

9mm.

seedcoat

endosperm

cotyledons

hypocotyl

radicle

O

A

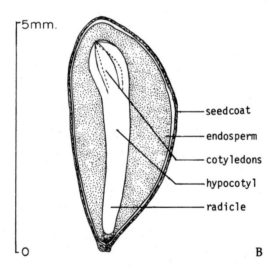

5mm.

seedcoat

endosperm

cotyledons

hypocotyl

radicle

O

B

Figure 1. Seeds of conifers. (A) Ponderosa Pine (*Pinus ponderosa*): longitudinal section through a seed, 6X. (B) Brewer Spruce (*Picea breweriana*): longitudinal section through the embryo of a seed, 12X. Note that the embryonic plant in a conifer seed is usually surrounded by a relatively large amount of endosperm. Both of these species of conifers have multiple cotyledons. (From, *Seeds of Woody Plants in the United States.* Handbook 450, Forest Service, U.S. Dept. Agric., Government Printing Office, Washington, D.C.

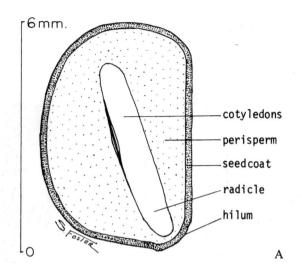

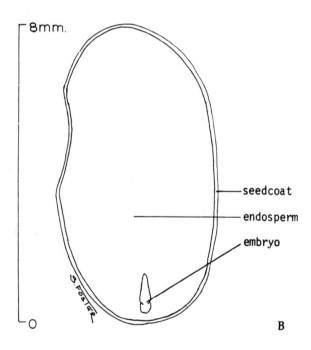

Figure 2. Seeds of monocotyledonous plants. (A) Soap Tree Yucca (*Yucca elata*): longitudinal section through embryo of a seed, 10X. (B) Fan Palm (*Washington fililera*): longitudinal section through the embryo of a seed, 10X. (From, *Seeds of Woody Plants in the United States.* Handbook 450, Forest Service, U.S. Dept. Agric., Government Printing Office, Washington, D.C.

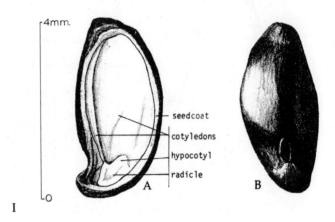

I

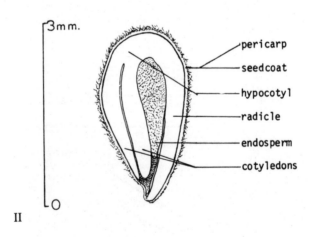

II

Figure 3. Seeds of dicotyledonous plants. (I) Roundleaf Serviceberry (*Amelanchier sanguinea*): A, longitudinal section through a seed, and B, exterior view; both at 12X. (II) Winterfat (*Ceratoides lanata*): longitudinal section through a utricle, 16X. (From, *Seeds of Woody Plants in the United States.* Handbook 450, Forest Service, U.S. Dept. Agric., Government Printing Office, Washington, D.C.

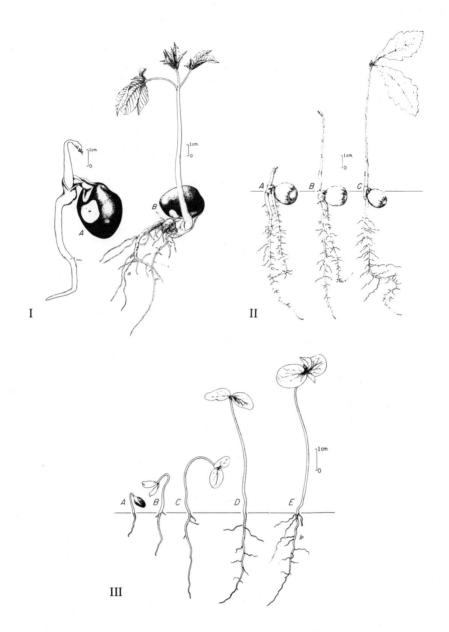

I   II

III

Figure 4. Types of germination. I and II hypogeal, cotyledons remain below surface of seed bed and III epigeal, cotyledons pulled through soil surface by hypocotyl arch, where they spread and act as leaves. (I) Yellow Buckeye (*Aesculus octandra*): seedling development at 2 and 4 days after germination. (II) Bur Oak (*Quercus macrocarpa*): seedling development at 1, 5, and 12 days after germination. (III) American Bittersweet (*Celastrus scandens*): seedling development at 1, 2, 5, 10, and 39 days after germination. (From, *Seeds of Woody Plants in the United States.* Handbook 450, Forest Service, U.S. Dept. Agric., Government Printing Office, Washington, D.C.

CHAPTER 3

Seed Collection

_____

TIMING COLLECTION

The timing of seed collection from wildland plant species is one of the most crucial and difficult steps in their propagation. Often the thought and desire to collect seeds of a wild plant is based on the sudden observation of a desirable species while hiking, fishing, or otherwise enjoying wildlands. Seldom do wildland plants in fruit catch would-be gardeners' attention. Exceptions are the cones of conifers and other showy fruits. Usually the flowers are what attract collectors' attention, and obviously most plants in full bloom do not have mature seeds. Collectors must plan ahead, mark the attractive plant in full bloom, and return when the seeds are mature.

Collection of immature seeds results in low seed viability or dormancy. The danger of delaying collection is that the fruits of many wildland plants dehisce (fall from the seedhead) very rapidly, so seeds are lost if collection is delayed. Collection of seeds from the soil surface may be possible, but it usually results in low-quality seeds and excessive cleaning cost.

Most crop plants bloom in sequence, beginning with the uppermost or central flower. Crop plants have been selected to have this type of flowering so the seeds will mature at the same time to aid in harvesting. These crop plants are said to have determinate inflorescenses or flower arrangements. In contrast, many wildland plants have indeterminate inflorescences where the flower stalk continues to grow with prolonged flowering and many different stages of seed maturity on the same flower stalk. This makes uniform seed collection difficult. If the seed collector is able to selectively harvest on the ripe portions of the seed stalk, the indeterminate inflorescence is no great problem.

A major factor that influences the collectors' approach to collecting seeds in the wild is: how many seeds do you want? If you are a commercial dealer with an unlimited market, obviously you want all the seeds you can get. A private gardener who only wants to try a new plant in the rock garden may only want a few seeds. How many seeds are enough for a new species? If you have no idea of what the germination of the plant's seeds will be, it is a good idea to have at least 5,000 seeds. In a section on germination, we will explain why these 5,000 seeds are necessary.

Sources of Information

There is no substitute for experience in judging when to collect seeds of wildland species. To start a collection program for a species with which you have no previous experience is difficult. Essentially, the collector must investigate the phenology, or sequence of events in the life history of the species to be collected. The stages in phenology in which the seed collector is interested include: (1) flowering, (2) seed formation, and (3) seed maturity. Guides to flowering can be obtained from regional floras (Table 1). More detailed information can be obtained from the few specialty manuals available or articles that deal with the life histories of individual native plants.

If you understand the biology of the species from which you are planning to collect seeds, it may be possible to partially predict the current year's seed production. This knowledge will pay additional dividends when the collector attempts to propagate the species. A good example of this type of knowledge is provided by the native shrub Bitterbrush (*Purshia tridentata*). Bitterbrush is a member of the rose family and is one of the most important browse species for big-game animals, especially mule deer. With showy flowers and a variety of growth forms, it is a potential wildland plant for ornamental plantings. We probably know more about the life history of Bitterbrush than any other native shrub. It has been studied on wildlands for 50 years. The seeds of this species that are collected from wildland stands for sale in the commercial seed trade constitute a several hundred thousand dollar trade annually.

Bitterbrush flowers on second-year twigs. The current year's twig growth elongates from 2–10 inches during the growing season, depending on the site potential and the condition of the shrub. The next growing season these twigs will support flowers. Older twigs on the Bitterbrush plants do not have flowers. Good Bitterbrush seed crops

14

generally follow years where average or better effective moisture produces stem elongation averaging at least three inches.

If you examine potential sites for Bitterbrush seed collection in the autumn and observe the current year's leader growth of the twigs to average 3 inches or more, then the potential exists for good flower production on the plants occupying the site the next spring. Note this is potential, not absolute seed production. Factors that might interfere with this flowering and seed production are: (1) biological, such as deer or cattle browsing the potential flower-supporting twigs during the winter and (2) physical, such as the occurrence of a killing frost during the critical spring flowering period. Obviously, the prediction of production from native plant species is a high-risk business.

Flowering is the first phenological stage of which the would-be seed collector must be cognizant. Flowering is obvious for many species with colorful petals, sepals, or bracts; but careful attention is required to note anthesis (shedding of pollen) with many grasses. After flowering, the sequence of phenology is as follows:

1. Soft-dough stage. This stage is indicated by the excretion of dough from seeds when squeezed between the thumb and forefinger. Seeds collected at this stage generally have low viability, if they will germinate at all.

2. Hard-dough stage. The hard-dough stage can be judged by biting the grain, once the dough stage is completed. In other words, if you cannot squash the seed between thumb and forefinger, try biting it. Once the seed is fully mature, it is usually too hard to bite. Seed collection should start with the transition from soft to hard dough. The time interval between soft and hard dough is a good indication of how soon to repeat the collection. With these first collections, the chance of obtaining plump, fully matured seeds can be increased by not stripping the seed from the plant, but rather by cutting considerable plant material and allowing the seeds to dry on the plant material. In most species this procedure will allow the seeds to reach full maturity. Care must be taken to insure that the mass of plant material dries uniformly and does not mold. The collection of additional plant material to help insure full maturity of the seeds is one of the most important techniques to develop in order to be a successful seed collector.

3. Maturity. Obviously, the goal of wildland seed collectors is to harvest mature seeds. Unfortunately, maturity and seed dehiscence may occur at the same time. To make sure some seeds will be obtained, repeated collections are necessary. The collections should extend from the latter part of the soft dough stage until all seeds are lost. Each collection must be clearly labeled with the collection date, location, species, and stage of phenology based on physical appearance. Descriptive notes on associated plant or site factors that may aid in reidentification of the stage of maturity are valuable. The seed collector must keep records.

Moisture Content of Seeds

For seed collectors who are more analytically inclined, seed moisture curves can be used to estimate the optimum date for harvest. Moisture is high in immature seeds, usually about 60 percent, but drops to about 10 percent as the plants mature. In a later section under seed storage, we will present methods of measuring seed moisture content. If repeated measurements are made over time, a seed maturity-moisture curve can be constructed. The seed-moisture curve for each species will show a characteristic shape because of differences in the slope, or drying rate. The rate of seed moisture change varies with climatic conditions, but averages about 3 percent per day during the seed maturity period. With above-average hot weather, the slope of the curve temporarily increases; whereas during cooler rainy weather, the curve flattens. For many crops, the seed maturity-moisture curves are known and are used as guides to harvesting seeds.

Seed maturity-moisture curves obviously require more effort than the ordinary gardener is willing to expend to obtain seeds of one or two species. However, the principle of seed maturity and decreasing seed moisture content should be understood by all wildland seed collectors. Commercial wildland seed collectors often judge seed maturity-moisture curves by the appearance of intact and squashed seeds. Objectivity can be given to these judgment determinations by taking actual seed moisture percentages.

Germination Tests and Seed Maturity Curves

Germination tests on each collection, made over a period of phenological development, provide the ultimate basis for judging the correct time of harvest. Careful records of the phenological stage of development at the time of seed collection must be kept because seed germination must be related back to these conditions. You cannot collect the seeds and immediately conduct the germination tests because the seeds of many species will not germinate when freshly harvested. Often a period for additional maturity, called afterripening, is required after harvest before the seeds will germinate. Therefore, germination tests are run on seeds from a succession of collection dates the winter following harvest, and the results of the tests related back to the phenological development of the species at the time the seeds were collected. This information is used to judge the timing of collection for the next season. Remember that optimum germination may occur in the most mature seeds, whereas optimum seed yield may occur at an earlier stage of maturity before seeds are lost by shattering.

Because of year-to-year variation in growing conditions, no method provides an absolutely accurate prediction of the specific date for seed collecting. There is no substitute for common sense, based on biological knowledge, to guide the collection of seeds of wildland plants.

Extending Collection Period

The period of optimum seed collection can be extended by starting collection at low elevations and following maturation upslope. The same procedure can be applied to species that produce tillers that mature later than the main inflorescence. There are some exceptions to the follow-the-seeds-up-the-slope rule. Fall-blooming species of Rabbitbrush (*Chrysothamnus* spp.) flower first at high elevations and last on lower slopes.

Often the seed collector can take advantage of micro-environmental differences at a given location to aid in collecting mature seeds. If seeds are immature on north-facing slopes, plants of the desired species growing on south slopes will generally be at a more advanced stage of maturity. Plants growing in swales or along drainage bottoms may produce more seeds than the same species on arid south slopes.

There is a danger in this practice. The Wyoming subspecies of Big Sagebrush (*Artemisia tridentata*) dominates large areas in the semi-arid portions of the Intermountain Area, but in average or drier years it seldom flowers. In drainage ditches alongside roadways passing through Wyoming Big Sagebrush stands, sagebrush plants are often found with abundant seed. It is tempting to assume the seeds are produced on Wyoming Big Sagebrush plants that took advantage of the extra environmental potential of the roadside ditch. Close examination often reveals the seeds are produced on plants of the Basin subspecies of Big Sagebrush, which out-competes Wyoming Big Sagebrush for the choice habitat.

Taking Advantage of Wildfires

Collectors of wildland seeds, especially seeds of herbaceous species, should learn to take advantage of plants growing in areas burned in wildfires. The reduced population density of plants that reinvaded burned areas, plus the availability of nutrients, makes for excellent seed production. In later years, the burned areas will often support stands of successional shrubs, making for good seed collection sites for these species.

Seed Caches

Rodents, birds, and insects, especially ants, are voracious collectors of some seeds. For some species (e.g., juniper berries and Pinyon Pine nuts), the seed collector must race the natural predators in order to obtain any seeds unless productive bagging or screening is used. Some seeds, especially conifers and Bitterbrush, can be obtained from rodent caches. Seeds of warm desert annuals have ant-attracting glands. Such seed can be recovered from the refuse dumps of ant nests. Some species of ants store the viable seeds in the nest, and only chaff is left on the soil surface. The droppings of many animals contain viable seeds or seeds that have improved germinability after passing through the digestive tract. The difficulty with any of these collection methods from caches or droppings is that the quantity of seeds obtained is small, and often contaminated with pathogens.

COLLECTION METHODS

Collection methods are largely hand methods because the desired wildland species do not grow in pure stands, and the topography often limits use of mechanical equipment.

Grass Species

The seeds of grasses can often be collected by stripping. The stripper may be the collector's fingers or mechanical fingers on a truck-mounted or towed implement. The process consists of allowing the grass stems (culms) to collect between the fingers and the seeds to be scraped from the terminal inflorescence as the stripper moves forward. A simple seed stripper made from sheet metal and a gallon can may be a valuable tool for hand stripping. The culms of the grass plant fit between the teeth of the stripper, and the inflorescences are pulled loose to drop into the container. In actual practice, it is never this simple. The collector using a tin can collector should wear gloves and be prepared to guide and stuff the stripped seeds into the container.

The seeds of tall grasses can be harvested with a homemade stripper attached to a light truck. The stripper attaches to the front bumper of the truck. It should be slightly flexible to absorb bumps, but constructed of strong-enough metal to prevent bending out of shape. Flail bars are necessary to knock seeds from the grass heads. The height of the flailing bars is important and must be adjusted for different-height grasses.

In dense stands of annual grasses, a garden rake can be used to strip the seeds of some species. For large-scale mechanical harvest, the seed stripper is a very inefficient way of collecting seeds. Despite the inefficiency of strippers, a number of native grass species cannot be harvested satisfactorily by any of the conventional mechanical means, such as field combines, making it necessary to strip. If wildland grass species occur in large-enough stands or on topography that permits use of mechanical equipment, it is far more efficient to use a header or a forage harvester to collect the material for threshing than to attempt to strip the seeds. Headers are machines that clip the plants just under the seed head. Seeds are cured in piles and later threshed.

For the novice seed collector, the seeds of Thurber's Needle Grass (*Stipa thurberiana*) provide good practice in how to collect and process seeds of a grass species. Thurber's Needle Grass is also a good example of a species in which it is advantageous to pull the culms to allow the seeds to fully mature.

Forage harvesters can be used to chop mature grass stands. To those unfamiliar with agricultural equipment, a forage harvester is similar to a giant rotary lawnmower that is normally used to harvest forage crops, and conveys the mature grasses into a wagon for transportation. The chopped material is either cured for later threshing, or the herbage and seeds are broadcast together at the time of planting. This can be a highly satisfactory technique in local areas where long-distance transportation is not involved. The herbage provides a mulch to help establish the desired seedlings. A word of warning: make sure the harvest material is free from weed seeds.

Many highly-specialized harvesters, such as pneumatic-type strippers, bluegrass cylinder strippers, and suction seed reclaimers, are used commercially in the crop seed industry. Grass fields can be repeatedly harvested during the same growing season by simple modification of a standard grain combine. The cutter bar is covered with a section of split tubing or pipe so the grass stems are not cut, but slide under the combine before the grass stems slide under the covered cutter bar. Extra-large bats on the combine reel swat the seed heads, knocking mature seeds into the combine. Immature seeds remain in the seed heads and pass under the combine. The speed of the rotation of the reels should be increased four or five times over normal operating speeds for this system to work. If you have never operated a grain combine, the above description will seem like nonsense. It was included to illustrate what can be done with existing farm equipment to harvest seeds of wildland species.

If large-scale collection of a suitable abundant wildland species is being contemplated, it may be worthwhile investigating some of the sophisticated equipment. If you are interested, a good place to start is the Equipment Development Committee maintained by the major public land management agencies. They have compiled a review of the available literature on high-production grass seed collectors. This publication provides domestic and foreign sources of small combines and grass strippers. It also lists research organizations active in research and development of grass seed collection equipment and has a reference section for pertinent literature.

Broadleaf Herbaceous Species

For the average collector interested in landscaping with native species, herbaceous broadleaf species are of special importance. The seeds of many herbaceous perennial species can be collected by holding a tray or box under the inflorescence while shaking or flailing the mature seeds into the receptacle. Fiber glass trays used in photographic darkrooms are excellent receptacles for this purpose. They are light, strong, and have a curled lip that can be conveniently gripped. These are available at most photo stores. An ordinary flat cake pan from the kitchen is a satisfactory substitute. In using the pan or tray, a good pair of gloves is the collector's best friend.

For very small herbaceous annuals, the simplest method of collection may be pulling the entire plant and bagging the material in paper sacks. Leave the bags open in a dry, well ventilated location; and the seeds will mature in place.

Herbaceous species with capsules or other fruits that explode present a special problem. A good example of this type of plant is the Common Oxalis (*Oxalis corniculata*), which becomes a terrible weed in greenhouses. When the slender capsules are ripe, they look like upright fingers of green bananas. If you touch these capsules at the right stage of maturity, you are treated to an explosion of surprising violence. This explosion distributes sticky seeds through the greenhouse. The only way to collect seeds of this type is to carefully collect the fruits while they are immature and allow them to ripen in mesh bags.

Collecting the entire plant is the only way to harvest seeds from spiny annuals such as Russian Thistle (*Salsola iberica*), where seeds are produced axillarly over all the plant. When the entire plant is harvested, care must be taken to insure that the material dries without molding. A large burlap bag, such as a wool sack, provides a convenient container for plants like Russian Thistle. You might wonder why anyone would desire to collect seeds of a weed such as Russian Thistle. Weed seeds are used by research scientists to develop new weed-control methods. Seed collectors should be aware of this small but potentially lucrative market. We use Russian Thistle to illustrate total plant collection of herbaceous species.

For herbaceous species that have spike-type inflorescences, pods can be stripped from the spike as with grasses. Lupines are a good example of a broadleaf plant where stripping is possible. Remember that pods of legumes such as lupines have a much higher moisture content than grass flower parts and must be given special care in drying. The pappused seeds (*achenes*) of many of the species of the sunflower family can be lightly brushed or swept into bags if the collector times seed fall perfectly. For large seeded species like Arrowleaf Balsam Root (*Balsamorhiza sagittata*), a 16-pound grocery bag makes a good-sized collector. Again, a good pair of gloves is necessary equipment. Some members of the sunflower family, especially those classed as thistles, have large heads or inflorescences that are subtended or surrounded below by armed bracts or spines. It is impossible to strip these heads, even with heavy gloves. The best policy with spiny thistles is to clip the heads with pruning shears and carefully drop them in a bag for later threshing.

Shrub Species

The seeds of many shrubby species can be collected by holding a tray or box under the outstretched branches while flailing the bushes with a stick or paddle, or by sweeping the arms across the upper branches to loosen the seeds, which then shower into the receptacle.

For collecting Bitterbrush seeds by hand, the Inyo tray was developed on the Inyo National Forest, located on the east side of the Sierra Nevada Mountains in California. It consists of an aluminum tray 20 inches long, 30 inches wide, and rounded at the bottom to a depth of 8 inches. A handle is inserted along the long axis. For limited collections, a cardboard box serves the same purpose as will baskets and canvas bags. A lightweight, 20-gallon barrel provides a ridged lip over which to bend shrub branches for removing fruits. This procedure is effective with spiny shrubs such as Desert Peach (*Prunus andersonii*), where the fruits must be physically stripped from the branches.

Collecting shrub seeds by catching them in containers requires very careful timing and cooperation from the weather. Strong winds can cause a crop of Bitterbrush seeds to fall to the ground in a very short period of time. Speed in collection of shrub seeds is of utmost importance. Canvas or plastic sheeting spread on the ground to collect seeds loosened from branches is of limited value because of the time and difficulty required to spread the sheeting under low branches and over rocks.

Shrubby species with explosive capsules, such as *Ceanothus,* must have the capsules stripped before maturity and must be ripened in mesh bags or on tarps to avoid seed loss.

Seeds of some semi-herbaceous shrubs, such as Four-wing Saltbush (*Atriplex canescens*), can be stripped by tractor-drawn seed strippers. Combines (combination of headers and thresher) have been used to harvest Winterfat (*Ceratoides lanata*) seeds. Field trials of vacuum harvesting, either vehicle-mounted or a backpack model, have shown promise for harvesting seeds of several shrubs.

Tree Species

Trees are usually of sufficient size to permit mowing grass, clearing brush, and otherwise enhancing the environment to aid in seed collection. Most of these elaborate preparations are confined to plantations or seed orchards. In these situations, it is feasible to spread canvas or plastic sheeting under the trees to expedite the gathering of fruits. A suspended net or fiber glass screen with a catching pocket at the bottom is useful for trapping light seeds of Ash (*Fraxinus*), Elm (*Ulmus*), and Mountain Mahogany (*Cercocarpus*) species as they are shaken or flailed from the crown.

Small fruits of trees are often stripped or picked by hand by collectors on ladders. Simple hand tools such as wire hooks to pull limbs closer, shaking poles, shears, flails, and rakes are valuable aids for collection of tree seeds. Large fruits or cones are often knocked or shaken off, and gathered from the ground beneath the trees.

Prompt collection of fallen fruits will reduce losses to fungi, insects, animals, and birds. Conifer cones exposed to high soil surface temperatures may open in several days and shed seeds. Acorns of some Oaks (*Quercus* spp.) and seeds of several other species may either dry out or germinate when left on the soil surface.

The obvious difference between trees and native plants of other growth forms is the height of trees. Many conifers such as the true Firs (*Abies* spp.) fruit on their upper branches. For mobility and handling ease, long extension ladders, picking platforms, or scaffolds have been mounted on trailers, trucks, or rough-terrain vehicles. Collection capabilities are materially improved, but the investment costs are substantial. Safe operation requires careful attention to positioning of the vehicle and to horizontal-reach limitation of the elevated ladder or boom. Higher crowns can be reached by using climbing irons or sectional ladders. Climbing irons are simple and easy, so the books say, but the spurs cause severe damage to some tree species.

Sometimes limbs, tops, or entire trees may be cut to collect seed. Efficiency of seed collection may be improved by removing the conebearing tops of Spruce (*Picea* spp.) and by clipping selected branches of Poplar (*Populus* spp.).

Cone collectors often visit logging shows to recover cones from recently felled trees. Care must be taken to make certain that the seeds were sufficiently ripe when felling occurred.

As is the case with all seed collection, the cones of conifers should be stored in bags with good aeration after harvest. Burlap bags placed on divided racks provide good temporary storage for cones.

SUGGESTED ADDITIONAL READING

Brandenberg, N. R. 1968. Bibliography of harvesting and processing forage-crop seed, 1949–1964. U.S. Dept. Agric., Agric. Res. Ser. ARS 42–135. 17 pp.

Dalrymple, R. L. 1984. Pickup grass seed stripper. *J. Range Manage.* 37:285–286.

Harmond, J. E., J. E. Smith, and J. K. Park. 1961. Harvesting the seeds of grasses and legumes, pp. 181–188. In *Seeds. The Yearbook of Agriculture.* U.S. Dept. Agric., Washington, D.C.

Larson, J. E. 1980. *Revegetation equipment catalog.* U.S. Dept. Agric. Forest Service, Missoula, MT. 197 pp.

McKenzie, D. W. 1977. Survey of high-production grass seed collectors. Project Record. U.S. Dept. Agric. Forest Service, Equipment Development Center, San Dimas, CA. 13 pp.

Nord, E. C. 1963. Bitterbrush seed harvesting: When, where, and how. *J. Range Manage.* 16:258–261.

Schneegas, E. R. and J. Graham. 1967. Bitterbrush seed collecting by machine or by hand. *J. Range Manage.* 20:99–102.

Table 1. Regional floras and other books useful for the seed collector.

Bailey, L. H. and E. Z. Bailey. 1976. *Hortus Third.* Macmillan Publ. Co., Inc. New York, N.Y. 1290 pp.

Cronquist, Arthur, A. H. Holmgren, N. H. Holmgren, J. L. Reveal, and P. K. Holmgren. 1977. *Intermountain Flora.* Columbia Univer. Press, N.Y., and N.Y. Bot. Garden. (Not all volumes are yet published).

Fernald, M. L. 1970. *Gray's Manual of Botany* (Eighth ed.). D. Van Nostrand Co., New York, N.Y. 1632 pp.

Hitchcock, A. S. 1950. 2nd ed. revised by Agnes Chase. *Manual of Grasses of the United States.* Misc. Publ. 200 U.S. Dept. Agric., Washington D.C. 1051 pp.

Hitchcock, C. L., Arthur Cronquist, Marion Ownbey, and J. W. Thompson. 1969. *Vascular Plants of the Pacific Northwest.* Univer. Washington Press, Seattle, Wa. (Five volumes).

Jepson, W. L. 1957. *A Manual of the Flowering Plants of California.* Univer. of Calif. Press, Berkeley, CA. 1238 pp.

Martin, A. C. and W. D. Barkley. 1961. *Seed Identification Manual.* Univer. of Calif. Press, Berkeley, CA. 221 pp.

McMinn, H. E. and Evelyn Maino. 1951. *An Illustrated Manual of Pacific Coast Trees.* Univer. of Calif. Press, Berkeley, Ca. 409 pp.

Munz, P. A. and D. D. Keck. 1959. *A California Flora.* Univer. of Calif. Press, Berkeley, Ca. 1681 pp.

Smith, J. R. and B. S. Smith. 1980. *The Prairie Garden.* Univer. of Wisconsin Press, Madison, Wi. 219 pp.

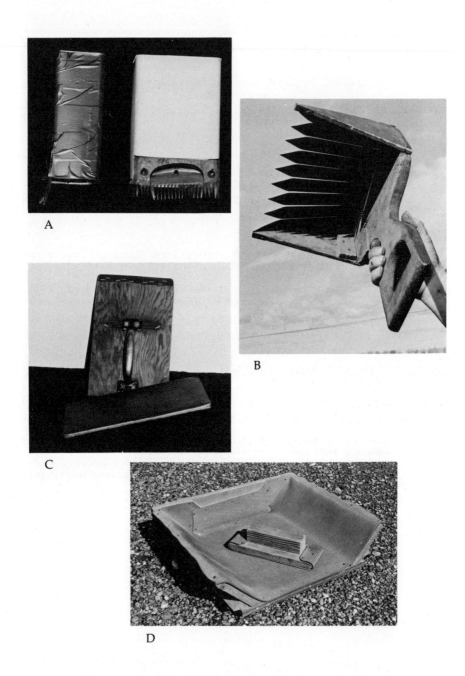

Figure 5. Seed collecting and threshing equipment. (A) Simple seed stripper made from gallon can and sheet metal, (B) Seed stripper made from wood and sheet metal, (C) Rubber-mat-covered paddles for threshing, and (D) Rubber-mat-lined threshing box and paddle. (From Young et al. 1981. *Collecting, Processing and Germinating Seeds from Western Wildland Plants.* U.S. Dept. Agric., ARM-W-3).

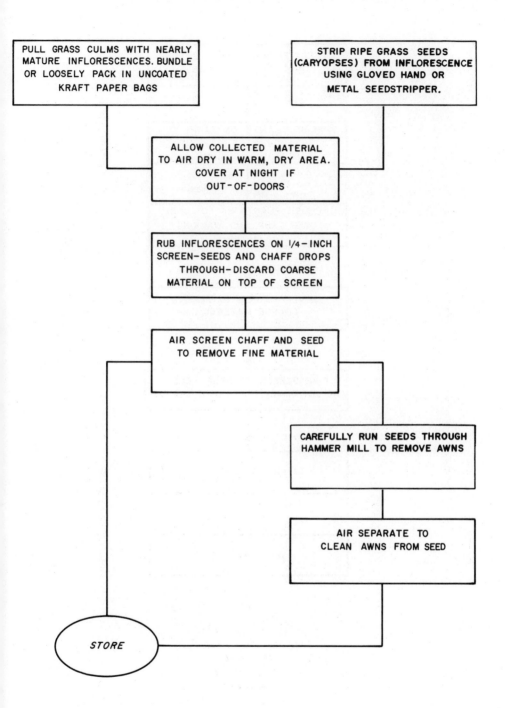

Figure 6. Steps in collecting, threshing, and cleaning seeds of Thurber's Needle Grass (*Stipa thurberiana*). (From Young et al. 1981. *Collecting, Processing and Germinating Seeds from Western Wildland Plants*. U.S. Dept. Agric., ARM-W-3).

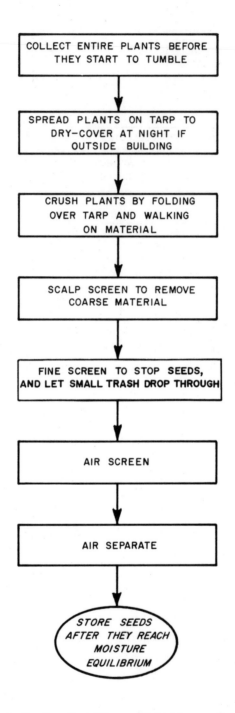

Figure 7. Steps in collecting, threshing, and cleaning seeds of Russian Thistle (*Salsola iberica*). (From Young et al. 1981. *Collecting, Processing and Germinating Seeds from Western Wildland Plants.* U.S. Dept. Agric., ARM-W-3).

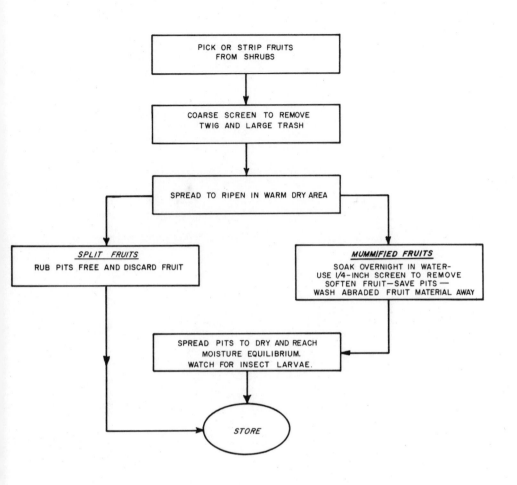

Figure 8. Steps in collecting and cleaning pits of Desert Peach (*Prunus andersonii*).
(From Young et al. 1981. *Collecting, Processing and Germinating Seeds from Western Wildland Plants*. U.S. Dept. Agric., ARM-W-3).

CHAPTER 4

Post-Harvest Handling of Seeds

Time is often allowed between seed collecting and threshing. This delay is necessary to allow fruits to mature and inflorescences to dry. Delays in threshing are also caused by the pressure for maximum collection during a short time period before the seeds fall from the plants and are lost.

This delay period is critical to seed yield and quality. We have stressed the use of mesh or paper bags for collection of seeds. Plastic or wax-coated paper should never be used. The moisture content of the freshly collected seeds is quite high, and plastic or other nonporous bags will trap this moisture and cause spoilage of the seeds.

A ventilated greenhouse room is an excellent post-harvest drying area for seeds to be threshed. If the collected material is dried outdoors, provisions must be made to cover the seeds at night in the event of winds, rain, or dew. If quantities of seeds are stored outdoors awaiting threshing, they are tempting targets for bird predation or infestation by rodents.

One method of reducing the amount of plant material to store and dry before threshing is to coarse-screen the collected seed or fruit in the field. This simple process consists of screening the collected material through a screen with large enough openings to allow the seeds or fruits to pass through. Coarse trash and waste are left on the top of the screen to be discarded. A second screening is made with a screen with openings too small to allow the desired seed to pass through. The seeds are kept on top, and fine waste passes through for discarding. Screens can be constructed from hardware cloth or fine wire mesh mounted on wooden frames.

The coarse screening is especially effective in removing relatively high-moisture-content coarse trash. In stripping or picking seeds or fruits, the collected material often is contaminated with leaves or small twigs. This is the type of material that should be removed in coarse screening.

THRESHING

Threshing of seeds from plants is an operation that predates civilization. In modern agriculture the threshing process is largely mechanized. Modern agronomic threshing machines are called combines because they are a combination of a feeding section (consisting of reel, divider, cutterbar, and feed mechanism) and a threshing section. In harvesting seeds of wildland species, we rarely have the opportunity to use a combine. Wildland plants do not normally grow in pure stands on level, rock-free topography. The processes inherent in the threshing section of a combine are worthwhile to review because these same processes are necessary in hand or mechanical threshing.

At the turn of the nineteenth to the twentieth century, threshing bees were major events in rural areas where cereal grains were grown. Stationary steam engines provided belt-driven power to threshing machines. Grain cut by Cyrus McCormick's reapers was hauled to the threshing machine on horse-drawn wagons. The stalks bearing cereal grains were pitched into the machine, where they wound around a cylinder and rubbed against concave bars. The turning cylinder rubbed the straw and seed heads against the fixed concave bars producing the threshing action. This action breaks the seed or fruit bases from the inflorescences and often removes the glumes, bracts, pods, or fruits covering the seeds. The adjustment of the clearance between cylinder and concave bars is of the utmost importance in mechanical threshing. A general rule is to have a clearance of one and one-half times the thickness of long seeds or one and one-half times the diameter of round seeds.

There are versatile small plot threshers that can be adapted to threshing small lots of seeds of wildland species. Special rubber-coated concave bars are available for these small threshers to reduce damage to fragile seeds.

HAND THRESHING

For most gardeners combining 100 bushels of grain per acre is a completely foreign thought. A much more frequent problem for them, and even for commercial collectors, is the threshing of a bag full of plant material. No matter how much material is threshed, the principles remain the same. Hand threshing must duplicate the action on concave bars in the mechanical threshing cylinders. The object is to rub the seeds to break loose the inflorescences or fruit covering.

An ancient method of threshing grain is to spread the material on a stone-floored courtyard or patio and walk oxen over the straw until their hoofs thresh the grains from the seed heads. Obviously, what is needed is a means of holding the plant material while applying abrasive action to free the seeds. A simple threshing method is to rub the collected material against a coarse screen with a gloved hand. A simple step up in technology involves rubbing the plant material on the screen with one or more paddles covered with rubber matting. Rubbing the collected material between the paddles or on top of a coarse screen threshes the seeds. If a large volume of material is to be threshed, this is a very time and energy-consuming operation.

A fully rubber-coated thresher can be simply constructed by cutting a discarded pneumatic tire tube and inserting the material to be threshed inside the tube. Threshing is accomplished by rolling the tube on the floor with your foot. Another simple threshing device involves the use of two clay bricks or the halves of a single brick. This technique is especially useful to seeds borne in capsules or non-splitting pods. Place the fruits, capsules, or pods between the bricks and press with a grinding action, leaving a few stems in the sample with the fruits to insure ample spacing so the seeds are not crushed.

MECHANICAL THRESHING

Hammer Mills

One of the first steps in mechanization is to use a hammer mill to rub the seeds loose. This is a most necessary operation for species where the fruit is so tough that hand rubbing is impossible. The novice seed processor must remember that hammer mills are designed to grind seeds for preparation of animal foodstuffs. Improper use of a hammer mill will destroy the ability of the seeds being processed to germinate.

The hammer mill utilizes many finger-like hammers rotating inside a section of perforated metal cylinder. Seeds processed in the mill are subjected to a vigorous beating or rolling action between hammers and perforated screen, which removes appendages and forces the seeds through the screen holes.

Results with the hammer mill depend on hammer speed, size of screen opening feed rate, and condition of the collected material. A good speed for the hammer mill in this procedure is about 50 percent of that used in normal grinding operations. If the speed is too fast or the screen too small, the seeds will be mutilated, cracked, or grated. The feed should be regulated so that the mill is approximately full at all times. If the mill is operated only partially full, seeds will be damaged.

Scalping

Once the seeds are rubbed free, they are rough-cleaned or scalped to remove the bulky foreign matter that will interfere with detailed cleaning. Stems, leaves, and trash not only interfere with fine cleaning but also contribute to high moisture content of temporarily stored samples.

SECOND-PHASE THRESHING

A threshing machine or combine has various mechanical cleaning or separating operations in line after the threshing cylinder. In processing small quantities of seeds of wildland species, the next operation after threshing by hand or hammer mill and scalping is usually seed cleaning.

HANDLING FLESHY FRUITS

Most of our discussion has applied to seeds that are dry at maturity or borne in dry caryopses or achenes. Fleshy fruits, including the berries, drupes, pomes, and those with seeds enclosed in fleshy arils, require special processing, which involves macerating the flesh, separating the seeds with copious use of water, drying, and cleaning. Processing should be started soon after collection to avoid damaging, fermentation, or mummification.

Mummification can be desirable. Seeds of species with a fleshy covering are sometimes dried and planted with skins intact. After initial cleaning or washing, such fruits may be spread on sheeting or in trays and dried in the sun. The mummified fruit may contain germination inhibitors and provide a nutrient-rich substrate for micro-organisms during germination.

Small lots of fruit can be macerated by hand. The flesh is hand-squeezed or mashed by a wooden block, rolling pin, or fruit press. Alternatively, flesh may be macerated by rubbing it against or through a screen. If the coarse screen is superimposed on a screen of a mesh sufficiently small to catch the seeds, a stream of water can be used to carry pulp away. Small-seeded fleshy fruits can be macerated in a blender. The addition of a small amount of detergent will help break the surface tension of the macerated fruit to produce cleaner seeds. Care must be taken to use a speed and duration of treatment that do not damage the seeds.

Machines suitable for processing large quantities of fleshy fruits include feed grinders, concrete mixers, hammer mills, and macerators. Most of these machines only free the flesh from the seed, and the residue must be separated from the seed by a later cleaning. When the pulp is largely washed away, the seeds and fruit skins can be dried and the skin subsequently removed in an air separator.

Following maceration and separation, careful drying is necessary to avoid damage to the recovered seeds.

CONIFER CONES

Seeds of most cones are released by drying cones to open them, shaking seeds out, separating seeds from cone scales and debris, loosening seed wings, and finally separating clean, full seeds from wings, dust, empty seeds, and other small particles.

As soon as the sacks of cones are brought in, they should be emptied. Immediate aeration is urgently needed when cones are wet and green. They can then be resacked loosely or spread on trays. The freshly collected cones should be sorted to remove foliage and debris before the cones open. Pitch is soft and sticky on freshly picked cones of many species. After drying, the chunks of pitch become loose and are difficult to remove from extracted seeds. A good practice is to allow a short period of drying to congeal the pitch, then wet and tumble the cones to remove both the pitch and dirt. Tumbling is done on a scalping screen. The key to this practice is to carry it out before the cones open. In drying freshly collected cones, allowance must be made for a two or three-fold expansion of cones as they open. For best aeration of drying cones, and uniform drying and opening, a layer one cone deep is recommended.

Prolonged cone storage of some species, 30 days or longer, may adversely affect seed viability. Storage of cones at temperatures near freezing is desirable for some species, but freezing temperatures should be avoided.

Given good drying conditions, cones of most conifers will open readily. However, serotinous cones of several Pines (*Pinus*) require pre-treatment and high heat, up to 170°F. Where weather is warm and dry during the extraction period, cones will fully open without artificial heat. Drying without direct exposure to sun is best for true Firs (*Abies*). Cones and seeds must be protected from rodents and birds.

Kiln-drying is used wherever natural drying conditions are not favorable for opening large quantities of cones. Small quantities may be dried in heated rooms or greenhouses. Cone kilns range in size from small cabinets to large buildings. They have in common a source of heat, a means of controlling movement of heated air by convection or forced draft, and some tray, shelf, or other system for exposing cones to moving air.

Small lots of cones can readily be dried by improvised means in a well-vented laboratory oven with circulating fan, over a hot-air register or radiator, or similar location. Whenever cones are dried, stringent fire precautions are necessary since dust, pitch, and dry cone material are highly flammable.

Tumblers of many sizes and shapes have been used for shaking out seeds from cones. Basically, a tumbler is a rectangular or round container mounted horizontally on its long axis, which turns at slow speed. As it turns, cones tumble about. Interior baffles often accentuate the jarring and tumbling action. Seeds fall from open cones through the high-strength wire mesh comprising the sides of the tumbler into a hopper.

Although some are loosened during tumbling and preliminary cleaning, wings must be removed from many conifer seeds. Seeds are dewinged by a variety of rubbing methods. Wings can be removed from small quantities by rubbing seeds between the hands or against a screen or roughened surface. The same principle is employed for larger quantities by gently tumbling dry or wetted seeds in a rotating container such as a cement mixer.

Scalping

Scalping of wildland seeds is usually done by hand screening. Again, we stress the principle of two screens, the first to let the seeds fall through and stop coarse trash, and the second to stop seeds and let fine trash pass through. For very small lots of seeds, kitchen strainers such as those used for loose tea serve very well for seed-cleaning screens.

Mechanical scalpers are made in many types but generally consist of perforated metal screens that turn on a central shaft and are inclined slightly from the horizontal. Materials fed into the higher end tumble inside the reel until the seeds drop through the perforations. Large trash stays in the reel.

Debearders

Many seed lots can be cleaned directly after scalping, but the awns of many grasses must be removed before the seeds can be cleaned. Awns or beards are elongated extensions of the glumes that subtend many grass flowers. The awns may be straight or bent, but are often sharp or barbed. Debearding machines are commonly used to remove these appendages and, in effect, to complete the threshing process.

The debearder consists of a horizontal beater that rotates inside a steel drum. The beater is made of a shaft with projecting arms, which are pitched to move the seed mass through the drums. Stationary posts protrude inward from the drum and prevent the mass from rotating with the beater. In operation, this machine vigorously rubs the seeds against the arms, post, and each other. The time the seeds remain in the unit is varied by regulating a weighted discharge gate. The severity of action is controlled by the exposure time, beater speed, and clearance between beater arms and post. Care must be taken not to damage the embryo portions of seeds. Debearding can also be done with a properly regulated hammer mill.

Once the seeds are threshed, the next step in processing is seed cleaning.

SUGGESTED ADDITIONAL READING

Harmond, J. E. and L. M. Klein. 1964. The versatile plot thresher. U.S. Dept. Agric., Agric. Res. Ser. Note ARS 42-4-1. 7 pp.

CHAPTER 5

Seed Cleaning and Separators

Cleaning of seeds collected from plants growing on wildlands is often a hand operation. However, mechanical seed cleaning equipment greatly facilitates cleaning of even small lots of seed, and sometimes is a necessary prerequisite for cleaning.

In agriculture, much of the seed cleaning is done in the field before the crop is harvested. Good weed-control practices minimize weed and other contaminant problems. In collecting seeds of wildland species, agronomic practices are seldom adaptable to the stand producing the seeds.

MECHANICAL SEED CLEANERS

The characteristics used in making separations of seeds and contaminants are size, shape, density, surface texture, terminal velocity, electrical conductivity, color, and resilience.

Many types of seed-cleaning machines are available that exploit these physical properties of seeds, either singly or in some combination. There are air-screen cleaners, specific gravity separators, pneumatic separators, velvet rolls, spirals, indent cylinders, indent disks, magnetic separators, electrostatic separators, vibrator separators, and others.

The choice of machines used and their arrangement in processing depends primarily on the seeds being cleaned, the contaminants in the mixture, and the purity requirements that must be met.

CHOICE OF MACHINES

The most widely used seed cleaning machine is the air-screen unit. It is the basic machine used in commercial seed cleaning. It makes separations mainly on the basis of size, shape, and density.

Air-screen Cleaners

There are many makes, sizes, and models of air-screen cleaners available. With wildland seeds, the volume of seed is normally so small that a two-screen cleaner is adequate. Typically, with a small, two-screen cleaner, seeds flow by gravity from a hopper into an airstream, which removes light, chaffy material so the remaining seeds can be distributed uniformly over the top screen as follows:

1. The top screen scalps or removes large material.
2. The second screen allows dirt and fine trash to pass, but retains seeds. The seeds are then passed through an airstream, which drops the plump, heavy seeds but lifts and blows light seeds and chaff into the trash bin.

Screen numbering system. The size of a round-hole screen is indicated by the diameter of its perforations. Perforations larger than 5.5/64ths of an inch are measured in 64ths. Therefore, a one-inch round-hole screen is called a No. 64, a 1/2-inch screen is a No. 32, and so on. Screens smaller than 5.5/64ths of an inch are measured in fractions of an inch (that is, 1/13, 1/14, and so on). The smallest round-hole perforation commonly used in air-screen machines is a 1/25 (0.04 inch).

Oblong-hole screens are measured in the same manner as round-hole screens except that two dimensions must be given. The hole width is indicated in 64ths of an inch; for example, 11 x 3/4 means openings 11/64ths of an inch wide and 3/4 of an inch long. In slotted screens smaller than 5.5/64 x 3/4, width is generally indicated in fractions of an inch, for example, 1/12 x 1/2.

Wire-mesh screens are designated according to the number of openings per inch in each direction. A 10 x 10 screen has 10 openings per inch across and 10 openings per inch down the screen. Such screens as 6 x 22 have openings that are rectangular in shape and are the wire-mesh equivalent of oblong-perforated or slotted screens. Triangular screens are usually measured by indicating the length of each side of the triangle in 64ths of an inch.

Selected screens. The two basic screens for cleaning round-shaped seeds are a round-hole top screen and a slotted bottom screen. The round-hole top screen should be selected so as to drop the round seeds through the smallest hole possible, and retain anything larger. The seeds drop through the top screen onto the slotted bottom screen, which takes advantage of the seed shape and retains the round, good seeds while allowing broken crop seeds and many weed seeds and dirt to pass through.

The basic screens for lens-shaped seeds are usually a slotted top screen and a round-hole bottom screen. The lens-shaped seeds tend to turn on edge and drop through a slotted screen but lie flat and travel over a round-hole bottom screen, which will allow most other crop and weed seeds to pass through.

An experienced air-screen operator can make unbelievable separations among seeds. If large amounts of seeds of wildland species are being collected, it may be worthwhile to get an experienced operator to air-screen the seeds. If no one is available or experienced with a given species, it will be necessary to experiment to find the correct selections.

If only handfuls of seeds of a given species are being collected, the screens from an air screen can be used for hand separations. At least one manufacturer of air-screens sells screens mounted in a hardwood frame for hand cleaning.

Adjustments. Besides the different types of screens used, the air-screen machines can be adjusted for rate of feed, airflow, oscillation of the screens, and screen pitch.

Generally, the grading or top screen should only be seven-eighths full. It is better to have a small section of the screen uncovered part of the time than to flood the screen occasionally.

Airflow is usually regulated by means of baffles in the air ducts. The airflow at the top of the duct is adjusted to blow out light, chaffy material and dust. Collections of seeds from wildlands that are threshed by hand are often very trashy. Seed collections of many wildland species will contain a high percentage of blank seeds where the fruit failed to fill. To remove these light seeds, the airflow at the bottom of the duct is regulated to a higher intensity to blow out all but the heavier seeds.

Oscillation of the screen is controlled by means of variable-speed pulleys and should be adjusted to keep the seed action "alive" over the screen. If too fast, the seeds hop across the screen and will not be separated. If too slow, the seeds have a tendency to slide. The pitch can be adjusted for each screen. The steeper the pitch, the faster the flow and the less time for separation.

Air Separator

Many different types of air separators are manufactured for seed processing. Some are called aspirators and others, pneumatic separators, but all use the movement of air to divide materials according to their terminal velocities.

Terminal velocity can be thought of as the velocity of a rising air current that will suspend a given particle. Although terminal velocity of a particle is a single characteristic, it is dependent on other properties—shape, size, surface texture, and density, all of which influence the particles' aerodynamic behavior in the airstream. If two different particles are fed into a rising airstream whose velocity is midway between the terminal velocities of the particles, one particle will rise and one will drop.

Two types of air separators are based on the location of the air movement unit. In the aspirator, a fan is placed at the discharge end of the air column, where it induces a partial vacuum so that atmospheric pressure can force air through the system. In the pneumatic separator, the fan is at the intake end of the air column, where it creates a pressure greater than atmospheric, which again forces air through the system. In both cases, air velocity through the machine can be adjusted by regulating the fan air intake. Seeds and contaminants with terminal velocities less than the air velocity through the units will be lifted. Materials with the same terminal velocity as the air will float, and objects with higher terminal velocities will fall against the airflow.

If only a small amount of seeds is to be cleaned, the air-screen may be bypassed in favor of a seed blower. The South Dakota Seed Blower, or some modification, is usually found in most seed laboratories. This very simple machine consists of a motor-driven fan to supply air pressure, and baffles in a vertical plastic cylinder for separation. Seeds are dropped into the vertical cylinder, and the air gate is opened until the airflow agitates seeds in the cylinder. Light chaff will float out the open end of the cylinder. Various grades of seeds, based on weight, will be trapped in the baffles on the side of the cylinder. Stones and heavy trash will remain in the bottom. Ideally, the airflow is adjusted so no viable seeds are lost. This is a quick, visible method of cleaning small lots of seeds. In contrast to these small batch-type machines, there are large continuous-airflow separators in commercial cleaning operations. An excellent air separator for hand threshing is a hand-held hair drier with the heat turned off.

Specific Gravity Separators

Specific gravity separators classify material according to density or specific gravity. Component parts of the gravity separator are an airblast fan, an air-equalizing chamber, a perforated deck, a variable-speed eccentric, deck rocker arms, a feeding or metering device, a deck adjustment, and a deck side-tilt adjustment. A mixture to be separated is metered at a uniform rate to the back of the deck. The slant of the deck and its oscillating motion move the seeds over the deck. Air forced up through the porous deck from the equalizing chamber forms thousands of small jets, which cause the material to stratify into layers of different densities. In the air-stratified material, the light material floats and the heavy material is in contact with the deck. The oscillating motion of the deck walks the heavy material uphill nearly parallel to the discharge edge, and the air floats the light material downhill. All the material travels from the feed point on the deck to the discharge edge; a gradation of material takes place, ranging from light material on the lower side of the deck to heavy material on the upper side. By means of movable splitters, the discharge can be divided into any number of density fractions.

The specific gravity separator's capability to make separations can be illustrated with a seed lot of Crimson Clover seeds contaminated with wild geranium seeds, dirt clods, and rocks. When the mixture is passed over a properly adjusted gravity separator, the geranium seeds and immature clover seeds will be discharged at the low side of the deck, the rock and heavy dirt clods at the upper side, and good crimson clover seeds in the middle.

The deck is the most important part of this machine. Various types of coverings are available to handle different types of seeds.

Other Seed Separators

If only a few hundred seeds of a plant species are to be cleaned, an efficient seed cleaner can be made from a sheet of newsprint. Fold the paper lengthwise and grasp each end at the folds. The seeds can be separated from the chaff and debris by gently inclining the paper while blowing on the material and juggling the paper. Although this may appear difficult, it is simple in practice to produce seeds sufficiently free of trash that only a final screening is required. The commercial seed industry has a host of other types of seed separators. Usually, the collector of seeds of wildland species will not have the need or the volume of seeds to require these specialized machines.

SUGGESTED ADDITIONAL READING

Harmond, J. E., N. R. Brandenberg, and L. M. Klein. 1968. Mechanical seed cleaning and handling. U.S. Dept. Agric., Agric. Handbook No. 354. Washington, D.C. 56 pp.

Mirov, N. T., and C. J. Kraebel. 1939. Collecting and handling seeds of wild plants. U.S. Dept. Agric., Civilian Conservation Corps., Forestry Publ. No. 5. 42 pp.

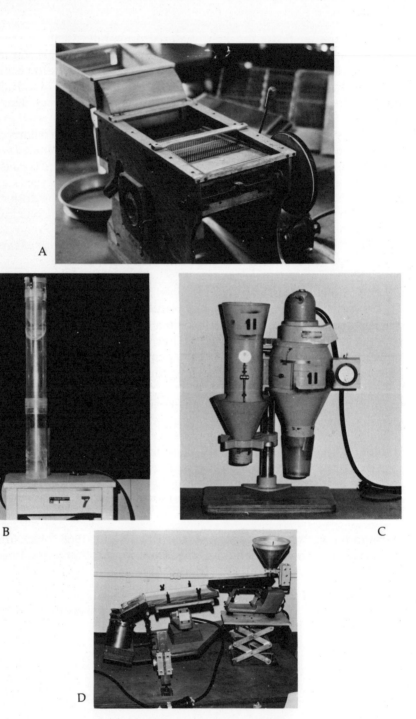

Figure 9. Seed cleaning equipment: (A) Airscreen, (B) Air separator—pneumatic, (C) Air separator—aspirator, (D) Specific gravity separator,

(E) Hammer mill, (F) Debearder, and (G) Belt separator. (Illustration from Young et al. 1981. *Collecting, Processing, and Germinating Seeds of Western Wildland Plants*. U.S. Dept. Agric. ARM-W-3).

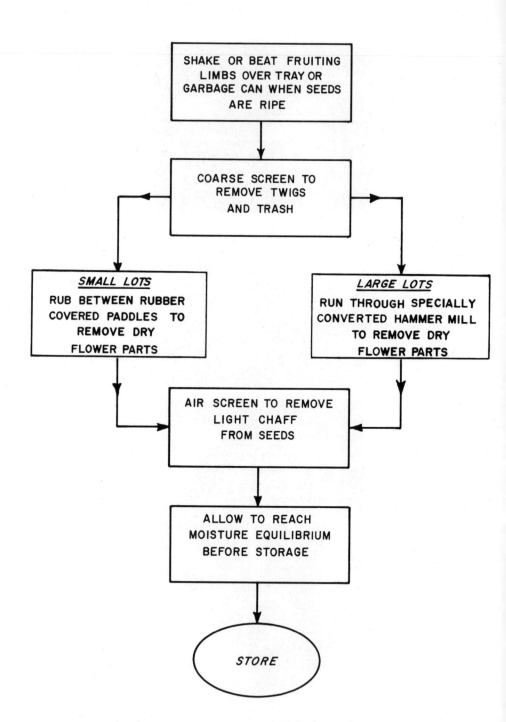

Figure 10. Steps in collecting, threshing, and storaging seeds of Bitterbrush (*Purshia tridentata*). (From Young et al. 1981. *Collecting, Processing and Germinating Seeds from Western Wildland Plants*. U.S. Dept. Agric., ARM-W-3).

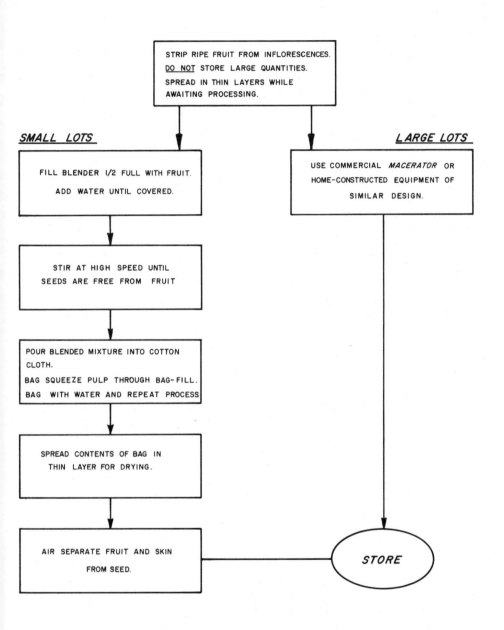

Figure 11. Steps in collecting, macerating, and cleaning seeds of Elderberry (*Sambucus glauca*). (From Young et al. 1981. *Collecting, Processing and Germinating Seeds from Western Wildland Plants*. U.S. Dept. Agric., ARM-W-3).

CHAPTER 6

Seed Storage

Even while on the plant, seeds deteriorate. High moisture, high temperature, sunlight, and diseases can adversely affect seeds before harvest. As previously mentioned, if the seeds are held at a high moisture content awaiting cleaning, decline in viability will occur.

IMPORTANCE OF MOISTURE AND TEMPERATURE

The two most important factors affecting seed longevity are seed moisture content and seed temperature. Two rules-of-thumb that relate the influence of the moisture and temperature on the rapidity of seed deterioration are:

1. Each 1 percent reduction in seed moisture doubles the life of the seeds.

2. Each 10°F reduction in seed temperature doubles the life of the seeds.

These rules give a quick insight into the importance of low seed moisture and low temperature in preserving high seed germination.

Moisture Content

The above rules-of-thumb must be qualified. If the seed moisture content is high enough (over 30 percent), nondormant seeds will germinate. From about 18 to about 30 percent, heating due to micro-organisms will occur if oxygen is present, resulting in rapid death of the seeds. From about 10 percent seed moisture for oily seeds or about 13–18 percent for starchy seeds, storage fungi grow actively and destroy the seed embryo. Therefore, seeds should be dried as quickly as possible to below 14 percent seed moisture and should be stored below this moisture content at all times. On the other hand, if seeds are dried below 4–5 percent seed moisture, deterioration is apparently somewhat faster than with 5–6 percent seed moisture.

Temperature

The rule-of-thumb for temperature is applicable down to at least 32°F. If seed moisture is below 14 percent, no ice crystals form below the temperature at which a seed freezes, so storage of dry seed at subfreezing temperatures should improve longevity. Unfortunately, most subfreezing stores have a high relative humidity, and after a period of storage the seeds gain moisture and ice crystals form. These crystals damage cells, causing loss in viability. If the seeds are first dried and then placed in moisture-proof containers, they will not regain moisture and should survive for a long period in subfreezing storage.

Relative Humidity

We have already mentioned relative humidity as influencing seed storage. This fact is so important in seed storage that it is worth expanding on the nature of moisture in the air environment of seed storage.

Relative humidity of the air is a percentage measure of moisture in relation to the total weight of moisture that can be held at a specific temperature when the atmosphere is saturated. As air temperature increases, the weight of moisture that a given weight of air can hold also increases. If the absolute weight of moisture remains constant, relative humidity decreases on heating. Cooling increases the relative humidity; and when it exceeds 100 percent, moisture will condense on the surface of the seeds. Since the moisture content of the seeds changes to remain in equilibrium with the ambient relative humidity, the seed moisture increases with cooling.

In a sealed container, there is a finite amount of moisture that can move into the seeds versus an infinite amount in open storage.

DRYING SEEDS

To dry seeds, the relative humidity of the air must be below equilibrium with seed moisture so there will be a moisture gradient from the seed to the air.

Lowering Relative Humidity

Seeds can be dried by unheated air only as long as the moisture gradient is from the seeds to the air. Unheated air will dry most freshly harvested seeds, but not to a safe moisture content. When the air is heated, the relative humidity is reduced and the moisture gradient from seed to air is increased; however, using air with too high a temperature can kill the seeds, especially if seed moisture is high. Even if the high-temperature drying does not immediately kill the seeds, it can cause injury resulting in loss of vigor and shortened storage life. A temperature of 100°F is a maximal for many seeds.

Drying can be hastened as follows:

1. By using high temperatures.

2. By increasing the airflow. About 0.3 cm cubic inches of air per cubic centimeter of seeds is the maximum economical airflow.

3. By allowing all sides of each seed to be exposed to the airflow as in a tumble drier or a baffled continuous-flow drier.

Too-rapid drying is dangerous. If the moisture gradient from the seed surface is steeper than the moisture gradient from the interior of the seed to the surface, the surface will dry rapidly, resulting in cracking of the tissue. For some seeds, too-rapid drying causes the outer cells to shrink and become impervious to moisture.

Removal of Moisture from the Air

The alternative to reducing the relative humidity in the air by heating is to remove moisture from the air, thereby reducing the relative humidity. This can be done by refrigeration or by dessication. Both of these methods have application to the commercial seed industry, but amateur collectors of wildland seeds seldom have such facilities available.

DRY STORAGE

To store properly, the seeds are dried first to the low moisture content desired. For one season's storage, the seed moisture content must be dried down to at least equilibrium with 65 percent relative humidity. Methods of measuring relative humidity are discussed in a later section. For 2–3 years of storage, seeds should be dried to equilibrium with 45 percent relative humidity. For long-term storage, after packaging in moisture-proof containers, the seeds should be dried to equilibrium with 25 percent relative humidity or 5–6 percent moisture content.

Maintaining Seeds in Dry Storage

After the seeds are dried to the desired moisture content, they must be kept at this level or the cost and benefit of the drying are lost. Maintaining the seeds in a dry condition can be done in three different ways, although the principles are the same in each method.

1. The storage itself is made moisture-proof and has dehumidification equipment to maintain the desired relative humidity.

2. The seeds may be packaged in moisture-proof containers.

3. The seeds may be placed in gasketed containers, enclosing also an indicator dessicant as well.

Moisture-proof storage rooms. There are specific standards for constructing moisture-proof storage rooms, but only the largest commercial wildland seed organizations can justify the cost involved in construction. If the individual collector of wildland seeds has access to existing facilities, he should take advantage of the close-to-ideal storage conditions.

Moisture-proof containers. Moisture-proof containers vary in form and capacity from aluminum foil laminated packets, tin cans, polyethylene (700-gauge) bags, and aluminum foil laminated drums to steel bins with gasketed lids.

The value of these containers as moisture-proof storage depends upon proper sealing.

Gasketed containers. For many small lots, which must be opened from time to time, a desirable moisture-proof container is a steel box with a gasketed lid. For every 10 pounds of seeds, 2 pounds of dry indicator silica gel is added in a cloth bag. If, on frequent opening, the indicator color changes from blue to pink, the silica gel should be replaced by a dry silica gel; and the replaced silica gel should be reactivated in an oven. In a cool room (70°F or lower), this method of storage will keep seeds of most species without loss of germination capacity for several years.

Problems Arising from Dry Storage of Seeds

Although dry seeds maintain viability better in storage, there are two cautions to remember. First, dry seeds are most easily damaged by the handling that will occur in transport and planting. Second, seedlings do not emerge as quickly from initially dry seeds as from seeds of higher moisture content. Dry seeds require more water and longer time to imbibe and germinate. Conditioning to equilibrium with 65 percent relative humidity will minimize transport damage and increase rate of emergence.

MEASUREMENT OF SEED MOISTURE

Since seed moisture is such an important aspect of the longevity of the seeds, accurate measurement of the moisture of a seed lot becomes important. The simplest moisture test, and one used for centuries, is the bite test. If the seed feels rubbery when bitten, the moisture content is dangerously high.

The most exact moisture test employs the Karl Fischer reagent. The reagent combines stoichiometrically (the methodology and technology by which the quantities of reactants and products in chemical reaction are determined) with water, and the amount of reagent combined is measured colorimetrically. All other methods of measuring moisture use the Karl Fischer method as the standard for comparison; however, this method requires a chemist, and is tedious and expensive.

Ovendrying Test

The International Seed Testing Association approves oven methods of moisture determination of either 220°F for 24 hours or 130° for 1 hour. This method can be accurate to ±0.1 percent seed moisture.

To use the ovendrying method you weigh the seeds, place them in the oven for the prescribed time, and reweigh them to determine the amount of moisture driven off by the heat. The weight of the dry seeds is divided by the original weight to determine the original percentage moisture content.

Infrared Balances

There are many types and models of infrared balances and electrical conductivity measures for determination of seed moisture. Some are portable enough for field use. Check manufacturers' standards for limits of accuracy, and be sure to calibrate against one of the standard methods for determining seed moisture content.

Electronic sensors are available for determining relative humidity. These are especially useful in small containers where standard psychrometers are difficult to operate. If the seeds have reached equilibrium with the relative humidity of the storage, measurement of relative humidity is a better measure of the storability of the seeds than seed moisture content. Many factors influencing seed storability, such as aging of seeds, fungal growth, and insect activity, are more directly correlated with relative humidity than seed moisture.

Additional Sources of Information on Seed Storage

Agriculture Handbook No. 506 by Oren L. Justice and Louis N. Bass is packed full of detailed information on the multitude of factors involved in the storage. This handbook is available from the Superintendent of Documents, U.S. Government Printing Office, Washington, D.C. 20402. Oren Justice is retired from the U.S. Department of Agriculture and Louis Bass is the Director of the National Seed Storage Laboratory, located at Fort Collins, Colorado. This illustrated handbook with a multitude of tables is a valuable addition to the library of serious seed collectors.

R. H. Ellis of the Agriculture and Horticulture Department of the University of Reading in the United Kingdom has prepared and updated a table providing storage characteristics for seeds of numerous plant species. This list has been published in the Plant Genetic Resources Newsletter (Bulletin 58) available from the FAO of the United Nations. Many native plant species from North America are included in this list.

SEED STORAGE INSECTS

Many insects that attack stored seeds were originally from the tropics. They spread and adopted to colder climates by living in man-made food storage shelters. Because stored-seed insects cannot remain active at low temperatures, their potential for development and damage is much greater in warm environments.

Most insect pests of stored seeds have a short period from egg to adult, their reproduction rate is high, and their adult lifespan is long. Temperature and moisture of the stored seeds influence the life cycles of the insect pest. Most stored-seed insects require temperatures of more than 50°F to develop damaging populations.

Stored-seed insects obtain water primarily from the seeds themselves. If the moisture content of the seed is low, generally less than 10 percent, the insects must obtain water by breaking down the grain components or by using their own energy reserves. Under these conditions, fewer insects survive. Properly applying these two natural factors—temperature and moisture—is fundamental to providing proper protection from storage insects.

When freshly harvested seeds are ready for storage, a few obvious safeguards can be employed. Do not store fresh seeds in contaminated containers. If small containers are used, cleaning before storage is simple. If large, bulk storage is used, cleaning is more difficult. Make sure no remnant seed from a previous year's storage is left in the bin in corners or cracks. Some insecticides are available for treating seed storage areas.

To help identify the kinds of pests that damage stored seeds, Agriculture Handbook No. 500, "Stored-Grain Insects," is available for sale from the Superintendent of Documents, U.S. Government Printing Office, Washington, D.C. 20402.

Once you discover that storage insects have contaminated your supply of wildland seeds, your choices of action are limited. Control of the insects probably requires fumigation. Because fumigant chemicals are highly toxic and hazardous to use, they are classified as restricted pesticides. Special training and certification are required before these materials can be purchased or applied. Persons desiring certification should contact their local Cooperative Extension Service specialists or State Board of Agriculture for further information about restricted-use pesticides. Because seeds of wildland species are not of major economic importance, there may not be a pesticide registered for use on the specific species contaminated. It is much better to avoid the problem with clean storage containers and proper moisture and temperatures for storage.

SUGGESTED ADDITIONAL READING

Ellis, R. H. 1984. Revised table of seed storage characteristics. Bulletin 58. *Plant Genetic Resources Newsletter.* Food & Agric. Organization of the United Nations, Rome, Italy.

Harrington, J. F. 1973. Problems of seed storage. In W. Heydecker, ed. *Seed Ecology.* Pennsylvania State Univ. Press, University Park, Pa. 578 pp.

Justice, O. L. and L. N. Bass. 1978. Principles and Practices of Seed Storage. Agricultural Handbook No. 506. Science and Education Administration, U.S. Dept. of Agric., Government Printing Office, Washington, D.C.

Leadem, C. L. and D. G. W. Edwards. 1984. A multiple-compartment tree seed tumble-drier. *Tree Planters Notes* 35C37:23–25.

U.S. Dept. of Agric. 1979. Stored-grain insects. U.S. Dept. Agric., Agric. Handbook No. 500. 54 pp.

CHAPTER 7

Seed Germination

Once the collector of wildland species seeds has mastered the problems of timing collections, threshing, cleaning, and storage, propagation is possible if the seeds will germinate. Some private and government laboratories will test the germination of submitted seed lots. These laboratories are largely geared to seeds of agronomic species and for conducting germination tests as prescribed by either national or international standards for testing seeds. Unfortunately, for most wildland species there are no standards for conducting germination tests. If the seeds of the desired wildland species are dormant, the collector may learn little more than that fact, unless a sympathetic seed technologist has the time to experiment with the seed lot.

If the collector wants to know if certain seeds will germinate or what kind of dormancy is limiting the germination, he or she is faced with solving the problem alone. Not all wildland seed collectors have the time or inclination for such endeavors, but a great deal can be learned from some rather simple tests.

GERMINATION TEST EQUIPMENT

Germination Environment

In order to germinate, seeds have to take up liquid moisture faster than they lose moisture vapor to the air. Most seeds require burial in the seedbed to accomplish this conservation of moisture. In seed testing, the same principle of moisture conservation applies. This is usually accomplished by placing the seeds in a closed container with a moisture-supplying substrate under the seeds. The container must be large enough to insure the atmosphere is not exhausted by the respiring seeds, but not so large that the atmosphere does not become rapidly saturated with moisture vapor. Obviously, the size of the container depends on the size and the number of seeds being germinated.

The standard container for germination tests has been the Petri dish. Originally, these consisted of about 4-inch-diameter glass plates, $\frac{1}{2}$-inch deep. One plate, the bottom, is made slightly smaller than the other, the top, so they overlap when nested together. Petri dishes are currently manufactured in a variety of plastic materials besides glass. Disposable plastic dishes are cheap and available from the biological supply houses. Many commercial seed laboratories have switched to plastic germination boxes, which are easier to stack and to fit with germination paper than circular Petri dishes. A variety of plastic kitchen containers can be adapted to serve as germination containers.

Besides a container, a substrate to maintain moisture is necessary. Blotter, filter paper, or commercial germination paper is used. Paper toweling may contain germicides which limit seed germination. Most paper toweling will not inhibit germination, but care should be taken in choosing a substrate. Besides paper, sterile sand, vermiculite, perlite, or mixtures of all three can serve as a germination substrate. Commercial seed-testing toweling or indented pads to hold a given number of seeds are available from firms that service grain elevator and seed-testing laboratories.

With seeds for which germination standards have been published, the type of substrate is specified. For example, it may call for testing of certain species seeds in sand or others on top of one piece of germination paper. For most wildland species, trial and error must suffice.

Germinators

The purpose of a germinator is to provide an environment conducive to seed germination. Generally, it is designed to provide a specific temperature with a free moisture-supplying substrate and a near-saturated relative humidity around the seeds. In addition, alternating diurnal temperature regimes or light of specific quality may be necessary. As we have already discussed, the saturated relative humidity and moisture-supplying substrate can be provided with Petri dishes and germination paper, or their equivalent. The use of dishes makes it necessary for the germinator to provide only the proper temperature for germination plus light, if necessary.

The most common type of available commercial germinators is designed to maintain a temperature higher than ambient current room temperature and a nearly saturated relative humidity from a free water surface. More expensive models of commercial germinators provide refrigeration for temperatures lower than the ambient. At additional expense, lights are available with automatic controls to provide diurnal temperature fluctuations and photoperiods.

Seeds of many wildland species germinate under cool to cold conditions, making a refrigerated germinator important. For the collector who cannot invest in an expensive germinator, an old refrigerator can be easily modified to serve the same purpose. Generally, pre-1960 refrigerators which function properly are better for this purpose than more modern models.

Temperatures can be varied in these old refrigerator germinators by adjusting the temperature control and choosing different shelf locations.

Using an inexpensive thermometer, it is relatively easy to calibrate the temperatures in various parts of an incubator constructed from an old refrigerator. Try the vegetable crisper for moderate temperatures and the meat storage area for colder temperatures.

The light bulb in the refrigerator should be removed or loosened so it does not function. Incandescent light from bulbs usually inhibits or reduces germination of light-sensitive seeds. Light from fluorescent tubes often enhances the germination of sensitive seeds. It is not difficult to construct a lighted germinator from an old refrigerator. A cool-white fluorescent light source is necessary for germination tests of seeds requiring light. A small fixture can be rigged inside the refrigerator. High light intensities are not required. An intensity of at least 4.6 lux (50 footcandles) and preferably 7–11.6 lux (75–125 footcandles) should be provided. Fluorescent tubes of 20–25 watts will suffice for a light source.

Fungicides

Healthy seeds being incubated at the correct temperature for germination often have natural resistance to micro-organisms. If the seeds become covered with fungal or bacterial colonies soon after they are put in the germination plates, it often is an indication of (1) non-viable seeds, (2) seeds with low vigor, or (3) a grossly inadequate germination environment.

If seeds are germinated at moderate to high incubation temperatures, micro-organisms often will colonize the germination substrate and seeds. Fungicides can be added, but it is usually better to avoid the problem by incubating the seeds at cooler temperatures or by careful sanitation. Make sure Petri dishes, germination pads, and working areas are clean. Seeds can be partially surface-sterilized by pouring boiling or near-boiling water over the seeds while they are held in a screen or by rinsing the seeds with a 0.01 percent solution of hydrogen peroxide. If a fungicide must be used, care must be taken in determining the proper concentration to insure germination is not lowered or inhibited. Groups of rolled germination blotters or Petri dishes may be kept in a large plastic bag, which will minimize infections and help maintain high humidity between inspections.

TESTING GERMINATION

Initial Test

Once seeds of the desired species are collected and cleaned, a germination test can be conducted. A surprisingly large number of collectors and novice native plant culturers find it difficult to take this initial step. Probably this is a result of the fear that seeds will be completely dormant.

Afterripening. The seeds of many species will not germinate, or have limited germination at maturity. Gradually, over time and rather independent of post-harvest storage conditions, this type of seeds become germinable. This post-harvest period of transitory dormancy is called an afterripening requirement. The length of this afterripening period varies greatly among species. For many species, one month or less after maturity is sufficient. A common duration for one type of after-ripening is three months. Under natural conditions in field seed beds, this three-month period protects the seeds that mature in mid to late summer from germination until fall rains assure seedling establishment and survival. The usual explanation of the afterripening mechanism is that the embryo within the seed is not mature when the seed falls from the mother plant. Embryo maturity gradually occurs during the afterripening period.

Several species of wildland plants have temperature-related afterripening requirements. Generally, seeds of this type will germinate at low temperatures but not at moderate or high temperatures until afterripening requirements are satisfied. A good example of this type of afterripening is provided by seeds of Russian Thistle (*Salsola iberica*). In the fall when Russian Thistles are mature, they will germinate at a very restricted range of moderately warm incubation temperatures. Gradually over the winter the afterripening requirement breaks down, so the seeds germinate at a wider range of temperature. By late spring Russian Thistle seeds will germinate at 32°–90°F or the improbable combination of freezing nighttime and 90°F daytime temperatures. This system of germination control fits well with the ecology of Russian Thistle seedlings, which are not frost-tolerant.

In order to know if seeds of particular species have afterripening requirements, it is necessary to make repeated germination tests starting soon after maturity, and to continue making them for six months.

Low incubation temperatures are more likely to produce germination than higher incubation temperatures for seeds with afterripening requirements. As previously mentioned, the low temperatures also reduce growth of micro-organisms.

While conducting initial germination tests to determine if after-ripening requirements exist, other observations can be made to aid in determining the mode of germination or dormancy.

Hard seed. One of these important observations is to determine if the seeds imbibe moisture. Imbibition is noted by observing swelling and softening of the incubated seeds. If the test seeds remain hard and dense, failure to germinate may be due to a hard seed coat that limits imbibition of moisture. This is especially common among species of small-seeded legumes.

If seeds fail to germinate at any temperature with three months' afterripening, or germination is severely limited, some form of germination enhancement will be required. If the seeds fail to imbibe moisture with initial testing, some form of scarification to make the seed coat permeable will be required. Hard seed coats do not break down in the transitory afterripening period.

Afterripening requirements are usually described as not being influenced by post-harvest storage conditions. In other words, you cannot do anything about the dormancy; you just have to let it wear off naturally.

Attempts have been made to accelerate the breaking down of after-ripening requirements by incubating the seeds dry at relatively high incubation temperatures, the theory being the high temperatures accelerate the maturity of the dormant embryo.

GERMINATION ENHANCEMENT

If the test species imbibes water but does not germinate, moist stratification is the next logical step. If the seed does not imbibe moisture, scarification is necessary. It is unfortunate that the words stratification and scarification bear a superficial resemblance and both apply to seed treatments. Stratification involves placing a seed in an environment where it will not germinate in order to promote germination when the seed is moved to a favorable environment. There is no physical damage to the seed coat. Stratification treatments are usually preceded by adjectives that identify the nature of the stratification environment, such as cool-moist, cool-dry, or warm-moist.

Seed scarification involves physical damage or removal of the seed coat. Scarification treatments are usually identified by the means of removing the seed coat, such as acid or mechanical scarification.

Seed scarification. The purpose of seed scarification is simply to break the impermeable seed coat and allow entrance of moisture. Scarification can be accomplished through mechanical abrasion or soaking in acid. Concentrated sulfuric acid (H_2SO_4) is commonly used to scarify seeds. The length of time required for scarification must be determined by experimentation. Length of treatment may vary from seconds to six hours, depending on the species being treated. Some particularly woody fruits may require 24-hour treatments in acid. Within narrower limits, the optimum duration of scarification treatment will vary among lots of seeds of the same species. After acid treatment, seeds should be neutralized and washed. The key to either mechanical or acid scarification is to avoid damage to the embryo. Remember, strong acids are dangerous and should be handled with care. Acid-proof gloves, clothing, and a face shield must be worn when working with strong acids. Always have a neutralizing solution prepared in advance of the treatment. Arrange for an eye-wash station at the location where the acid scarification is conducted. Use the smallest amount of acid that will cover the seed. Treat only small amounts of seeds at one time.

Many years ago N. T. Mirov designed a simple vessel for use in acid scarification of seeds. It consists of a stainless steel screen dish that can be lowered into a quart glass jar by a center pole. A small amount of acid is placed in the bottom of the jar, and the seeds to be treated are placed in the basket and lowered into the acid. After the prescribed treatment, the basket is raised and the acid allowed to drain before the basket is transferred to the neutralizing solution.

Some seeds are very sensitive to acid scarification. The acid does not just remove the seed coat. The process generates a lot of heat and so drives moisture from the seed. This can be partially counteracted by pre-chilling the acid before treatment and cooling the seed/acid mixture during treatment by immersing the treatment vessel in a water bath. After treatment, rinse a small volume of seeds and acid in a large volume of water. This also helps to dissipate the heat from the acid/water reaction.

For very small lots of seed, careful rubbing on fine sandpaper may be sufficient to abrade the hard seed coat. Mechanically threshed and cleaned seeds usually require less scarification than hard-threshed seeds of the same species because of the damage inherent in the mechanical handling of seeds.

The maternal environment in which seeds are produced influences the number of hard seeds. If a hard-seeded wildland species is grown under irrigation in a favorable environment, the number of hard seeds may be reduced.

Seed stratification. If the test seed imbibes water either before or after scarification but fails to germinate, stratification is a logical next step. The principles of stratification were developed by foresters. It involves pre-chilling conifer seeds in damp peat moss, with the seeds and moss arranged in large, flat trays stacked in refrigerators. The stacked trays apparently suggested a stratigraphic sequence, so the term stratification has become attached to the process.

Cold-moist stratification is the type most commonly used. The embryo of many seeds fails to germinate because oxygen does not diffuse through the seed coat. At cold temperatures, more oxygen is soluble in water, so the oxygen requirements of the embryo are better satisfied. Cold-moist stratification imitates overwintering under snow cover in a field seedbed.

Stratification requirements for seeds vary greatly among species, both in optimum temperature and duration. Generally, temperatures in the range from 34°–40°F are best. Time required for stratification varies from two weeks to several months.

One should not assume that a one-month stratification requirement will be satisfied by fall planting. The stratification requirement will be met only if temperatures are not excessively cold (not below freezing), if the seeds remain moist but not saturated, and if the cool-moist period lasts for a month without interruption. In many regions the cool-moist period occurs in the spring, when the seeds should have already germinated in order to survive summer drought. On the other hand, stratified seeds cannot be seeded in the fall without risking germination and frost damage.

Plastic bags are useful in stratification. The seeds are placed in the bags with moist sand or vermiculite and stored at low temperatures (33°, 40°, or 50°F), usually until the stratification requirement is satisfied. The seeds are then placed in cloth bags to recover from stratification. A variation of this method is to fill the plastic stratification bags with activated charcoal. Activated charcoal stratification is the only effective method for some species.

A variation of the stratification procedure is to stratify seeds in a cold-water bath where the oxygen content of the liquid is kept near saturation by forcing compressed air through the liquid. This method has given spectacular results with seeds of species that normally do not respond or respond poorly to stratification. The specific length of stratification must be determined by trial and error.

Besides cool-moist stratification, seeds of some species require warm-moist stratification or a combination of warm-moist followed by cool-moist stratification. Such combinations of requirements are very difficult to determine without considerable experimentation.

Occasionally, seeds require a very specific stratification temperature, such as 35°F; neither 32°F nor 40°F will satisfy the stratification requirement. Again, this type of requirement necessitates a great deal of experimentation.

It is important to remember that all stratification methods involve moist seeds. Sometimes, it is advantageous to allow seeds to imbibe water at room temperature before being placed in the stratification treatment. Dry seeds rarely demonstrate enhanced germination as a result of pre-chilling.

Widely fluctuating temperatures. Germination of seeds of certain species is enhanced when incubated at widely fluctuating temperatures. Inland Salt Grass (*Distichlis stricta*) is a good example of this type of response. Salt Grass seeds are usually nearly completely dormant when incubated at constant temperatures, and do not respond to any of the standard germination enhancement techniques. But when incubated for 8 hours at 105°F and then switched to 40°F for 16 hours daily, considerable germination results.

Moist heat. A sometimes successful alternative to scarification of hard seeds and an occasionally valuable enhancement treatment for seeds that do not respond to stratification involves immersion in hot water. Seeds are dropped into boiling water, the container is then immediately removed from the heat, and the seeds allowed to steep in the hot but cooling water. Duration of effective treatment may be from a few seconds to two hours, depending on the seed or fruit. The seeds of several species of *Ceanothus* respond to hot water treatments.

Water-soluble inhibitors. Soluble inhibitors can be washed from seeds with running tap water. Water cold enough to inhibit germination should be used in that after the leaching process the seeds can be planted without any damage to partially emerged radicles.

Water-soluble germination inhibitors can be removed by planting seeds in containers of damp sand and placing the containers on mist benches. The amount and incidence of mist should not be excessive but must keep the sand damp. For some species, especially some grasses, the germination inhibitor can be physically removed by removing awns or other appendages of the caryopses. Remnant flower parts on achenes may contain germination inhibitors, as in the case of Bitterbrush (*Purshia tridentata*). Relatively simple mechanical processing eliminates this type of germination inhibitor once the cause of dormancy is recognized.

Soluble inhibitors are also absorbed by activated charcoal.

Non-water-soluble inhibitors. Germination inhibitors that cannot be removed with washing or boiling can be removed with organic solvents. Soaking in acetone or methyl chloride enhances germination of some dormant species seeds. This type of treatment should not be construed as simply washing away an inhibitor. Chemical treatments probably involve complex reactions that alter permeability of seed coats or the biochemical status of the embryo itself. These and other chemical treatments for germination enhancement may influence oxidative processes in seed germination.

Hydrogen peroxide. The germination of seeds of several species, especially members of the rose family, is enhanced by soaking in peroxide (H_2O_2) solutions. Dramatic germination enhancement has been obtained with seeds of Bitterbrush (*Purshia tridentata*) and Currleaf Mountain Mahogany (*Cerocarpus ledifolius*). A wide range of concentrations from 1–30 percent is effective. Generally, the higher the concentration, the shorter the soaking time, but the greater the risk of damaging the seed. Hydrogen peroxide is a highly reactive chemical. Concentrations greater than 3 percent are particularly dangerous to handle.

Light. If after washing, stratification, scarification, and boiling, seeds still fail to germinate, the effect of light intensity, duration, and quality should be investigated. In practical seed testing sequence, the seed lot should be subdivided at the beginning, and all of the above procedures conducted at the same time. In some cases, a sufficient quantity of seeds of wildland species may be lacking, so a stepwise procedure may be followed to conserve seeds in hopes that the key to germination will be found before all the steps in the experimental procedure are completed.

To determine if seeds have a positive germination response to light, imbibed seeds should be exposed to cool-white fluorescent light for 8 in 24 hours. As previously mentioned, illumination of 75–125 footcandles is adequate. Occasionally, longer diurnal photoperiods may be more effective in promoting germination.

Seeds of many species germinate after exposure to red light but are inhibited by exposure to far red light. Germination enhancement for such seeds requires light filters, darkrooms, and light-proof containers. Generally, incandescent light should be avoided as seeds respond better to fluorescent light.

Sulphydryl compounds. Many sulphydryl compounds markedly stimulate germination of dormant seeds. The best known agent of this group is thiourea. The classical example of the use of thiourea on Western wildland species is Bitterbrush seeds. The seeds are soaked in a 3 percent thiourea solution, then dried before rewetting for germination. Thiourea is a very dangerous chemical and can be hazardous if used improperly.

Ethylene. Ethylene has long been known to elicit a wide range of physiological responses in plants. For numerous species, ethylene can be used to enhance germination as a gas or with the chemical ethephon ([2-chlorethyl] phosphonic acid), which produces ethylene when added to the germination substrate.

SECONDARY GERMINATION ENHANCEMENT

We have not discussed all of the treatments used to enhance germination of dormant seeds, but have reviewed the usual and most successful treatments. Frequently the primary treatments result in germination of some previously completely dormant seeds, but this level is so low that successful propagation is still quite limited. So other treatments must be utilized to enhance germination in combination with primary dormancy-breaking treatments.

Potassium nitrate. The germination of seeds of many species is influenced by the presence of nitrate ions. Potassium nitrate (KNO_3) is widely used as a source of nitrate ion enrichment. The addition of potassium nitrate should probably be ranked as a primary method of germination enhancement, but often is used to augment germination after scarification, stratification, or light treatment. Customary germination standards for seeds of many agronomic species require the addition of a 0.2 percent solution of potassium nitrate to the substrate during germination or as a part of pregermination chilling treatments. Lower concentrations of potassium nitrate are more effective for some species.

Gibberellic acid. The mode of action of gibberellic acid in seed germination is not known, but very low concentrations of this growth regulator can greatly enhance germination. Concentrations of from 1–150 parts per million (p/m) are commonly used in germination enhancement. Combinations of gibberellic acid and potassium nitrate are often more effective than either material alone.

A good balance is needed for preparing the minute concentrations of gibberellic acid used. A solution with a concentration of 1 p/m of gibberellic acid consists of 0.001 grams of gibberellic acid dissolved in 1,000 milliliters of water. Gibberellic acid is sold as a 10 percent active-ingredient preparation, which makes the weighing simpler. One alternative is to prepare higher concentrations than needed and dilute to the desired concentration. For example, 1,000 p/m would be 1 g in 1,000 ml; however, gibberellic acid is relatively expensive and breaks down very rapidly at warm temperatures.

Again, we have not exhausted all of the chemicals that have been used in secondary treatments to promote germination, only those proven to be most useful.

SUGGESTED ADDITIONAL READING

Chan, F. J., R. W. Harris, and A. T. Leiser. 1977. Direct seeding of woody plants in the landscape. Div. of Agric. Sci., Univ. of Calif. Leaflet No. 2577. 13 pp.

Copeland, L. O. 1976. *Principles of seed science and technology.* Burgess Publ. Co., Minneapolis, Mn. 368 pp.

Emery, D. 1964. Seed propagation of native California plants. Santa Barbara Botanical Garden Leaflet 1(10):81–96.

Hartman, J. T. and D. E. Kester. 1968. Sexual propagation. Pp. 53–188. In *Plant propagation—principles and practices.* Prentice Hall, Englewood Cliffs, N. J.

Heydecker, W. (editor). 1973. *Seed ecology.* Penn. State Univ. Press, University Park, Pa. 578 pp.

Maguire, J. D. and A. Overland. 1959. Laboratory germination of seeds of weedy and native plants. Wash. Agric. Exper. Sta. Cir. No. 349. 15 pp.

Mirov, N. T. and C. J. Kraebel. 1939. Collecting and handling seeds of wild plants. U.S. Dept. Agric. Civilian Conservation Corps. Forestry Publ. No. 5. 42 pp.

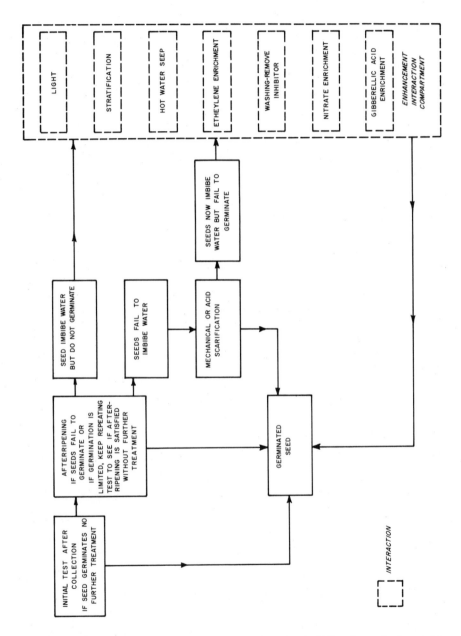

Figure 12. Steps in enhancing germination of seeds of wildland species for which no germination standards exist. (From Young et al. 1981. *Collecting, Processing and Germinating Seeds from Western Wildland Plants.* U.S. Dept. Agric., ARM-W-3.)

CHAPTER 8

Seed Technology

In recent years the seeds of many native plant species growing on wildlands have become important in the commercial seed trade. There are various legal requirements and procedures governing the industry that should be understood before entering into the seed trade.

ENDANGERED AND THREATENED PLANT SPECIES

Regulations governing interstate and foreign import and export of endangered and threatened plant species became effective July 25, 1977. The regulations establish a flexible system to allow legitimate commerce to continue in jeopardized plant species while protecting those plants remaining in the wild. Under the Endangered Species Act, it is illegal—except by permit—to import or export, to sell, offer for sale, deliver, receive, carry, transport, or ship in interstate or foreign commerce, listed plants. To obtain copies of the regulations and the up-to-date list of species, contact the Federal Wildlife Permit Office, U.S. Fish and Wildlife Service, Washington, D.C. 20240.

The Office of Endangered Species of the U. S. Fish and Wildlife Service is a good source for the names of publications with local and regional floras of endangered, threatened, and rare vascular plants. Check with the Fish and Wildlife Service regional offices in Honolulu, HI; Portland, OR; Denver, CO; Albuquerque, NM; Twin Cities, MN;

Boston, MA; and Atlanta, GA. The plant taxonomist at your local university is another good source for rare plant information.

In some states, there are laws and regulations governing intrastate commerce in native plants. The important point is to check to make sure the species you are collecting are not rare or endangered; or if you want to work with the rare and endangered species, obtain the required permits and follow the regulations.

LABELING SEED LOTS

The information required on a seed tag varies from state to state, but a generalized label requires the following:

1. Common name of species and subspecies or variety, if appropriate.
2. Scientific name of the genus, species, and subspecies or variety, if appropriate.
3. Lot number or other identification.
4. The origin of the seeds. For seeds collected from predominantly indigenous stands, information regarding the area of collecting by latitude and longitude, geographic description, or political subdivision such as state and county is required.
5. The upper and lower limits of elevations within which the seeds were collected.
6. The purity of the seed lot as a percentage of pure seeds by weight.
7. For species for which standard germination testing procedures are prescribed by state regulations, the germination percentage, percentage of firm ungerminated seeds, and the date of the test by month and year are required. For freshly harvested seeds, the label may say "test in progress."

   For seeds being transported to a consumer, the label may say "contract seed not for resale and subjected to test to be arranged."

   For species for which standard germination testing procedures have not been prescribed, only the calendar year in which the seed collected is required.

   Unfortunately, most wildland species do not have prescribed methods established for seed germination.
8. The name and address of the person who labeledthe seeds or who sells or offers them for sale.
9. For specific label requirements, check with the State Department of Agriculture of the state to which the seeds are being shipped.

Obviously, the label on a bag of seeds can be a very informative document. Considering the very high cost of the seeds of many wildland species, it is important to insist on properly labeled seeds and to read the label before purchasing.

Purity Testing

Seed purity denotes the composition of a particular seed lot. It is based on a physical determination of the components present and includes percentages by weight of: (1) pure seed, (2) other crop seed, (3) weed seed, and (4) inert matter. Pure seed is the portion of the working sample represented by the species for which that lot is being tested. Other crop seed is the percentage of crop seed, other than the species being tested, present in concentrations of less than 5 percent. Weed seed indicates the percentage of seeds present from plants considered weeds. Sometimes this designation may be quite arbitrary, since a plant may be considered a crop in one state or country, but termed a weed elsewhere. Inert matter denotes the portion of the sample that is not seed. It usually consists of chaff, stems, and small stones, but may include pieces of broken, damaged, or immature crop or weed seeds that do not qualify as entire seeds.

Noxious Weed Seed Examination

Each state has established an official list of noxious weed seeds. In general, the plants from these seeds are particularly troublesome and objectionable. Such lists are a part of the state seed law and are usually defined in two categories, primary (or prohibited) and secondary (or restricted) noxious weed seeds.

Sale of seed lots containing primary noxious weed seeds is usually prohibited, while the sale of seed lots containing secondary noxious weed seeds is permitted, but the number of weed seeds is limited. Since each state has its own seed law, the weed seeds listed as noxious are not necessarily the same from state to state.

Germination Testing

Probably the single most convincing and accepted index of seed quality is the ability to germinate. We have previously discussed the processes of germination and germination enhancement of dormant seeds. This discussion will cover only the laboratory techniques used for performing the analysis.

The germination test is ordinarily performed on the pure seed of the crop seed being tested after all inert matter, other crop seed, and weed seed are removed.

A minimum sample of 400 seeds is recommended for a statistically dependable germination test. These are usually arranged in four replicates of 100 seeds each.

The exact procedures and regimes under which seeds of different species are germinated have been developed out of over 75 years of experience in germination testing and have been augmented during the last 40 years by a systematic program of referee testing involving interchange of samples and results among laboratories.

The testing instructions given in the rules for seed testing include the germination media (substrate), the temperature required, the duration of the test period, and additional suggestions for optimal results. Unfortunately, these standards have not been developed for most plant species native to wildlands.

The time required for germination test varies among species. Some species require less than seven days, while others require a month or longer.

At the end of the prescribed germination period, the tests are evaluated. The seed analyst must determine if the seed has germinated. The usual definition of germination involves the normal seedling concept: "The emergence and development from the seed embryo of these essential structures which, for the kind of seed in question, are indicative of the ability to produce a normal plant under favorable conditions."

Any seedling that is not classified as a normal seedling is considered abnormal. The germination analyst may classify seedlings as abnormal for various reasons, with the absence of essential structures being most common.

Seeds other than hard seeds that remain firm (nondecayed) and ungerminated at the end of the prescribed germination period are called firm, ungerminated seeds. Hard seeds are those which do not imbibe water and therefore remain hard at the end of the prescribed germination period.

Viability Test

Seed analysts have searched for a rapid and reliable test for seed viability that would eliminate the long germination test sequence for dormant seeds. Many methods of estimating viability have been examined, but embryo excision and tetrazolium staining are generally used and recognized in official seed-testing procedures.

Embryo excision. Seeds are cut open, and those with fully grown, firm, undamaged, healthy-looking tissue with the proper color are judged viable. Judged as nonviable are seeds with milky, unfirm, moldy, decayed, shriveled, or rancid-smelling embryos and abortive seeds that have no embryo. The problem with this test is that extremely small or very vitreous seeds are impossible to cut without crushing the embryo.

Tetrazolium test. Biochemical staining tests have been developed that visibly stain viable seeds. The commonly used test is the tetrazolium method. The Association of Official Seed Analysts has a handbook that should be followed when conducting tetrazolium tests.

Tetrazolium tests are only definitive when standards have been developed for interpreting the results. Just because seed tissue stains red with tetrazolium does not mean the seed will germinate when factors causing dormancy are overcome.

SUGGESTED ADDITIONAL READING

Copeland, L. O. 1976. *Principles of seed science and technology.* Burgess Publ. Co., Minneapolis, MN. 368 pp.

Grabe, D. F. (editor). 1970. Tetrazolium testing handbook. Contribution No. 29 to the Handbook on Seed Testing. Association of Official Seed Analysts. 62 pp.

CHAPTER 9

Germination of Seeds of Trees

Gardeners can find more specific information about the germination of seeds of native tree species than any other growth form of wildland plants. This knowledge base exists thanks to efforts to reforest areas with direct-seeded or transplanted tree species. The Forest Service of the U.S. Department of Agriculture compiled and published the existing information on collecting and germinating seeds of woody plants in a handbook in 1948. This handbook was revised in 1974.

CONIFERS

The conifers or cone-bearing plants compose one of the major divisions of the Plant Kingdom. Characteristically, they are woody plants that usually have scaly to needle-like leaves. The seeds of this group are not enclosed in a seed vessel but are released when the cone scales pull apart (gymnosperm).

Abies—True Firs

Abies is a Northern Hemisphere genus of about 40 evergreen tree species. A connoisseur of the natural beauty of trees cannot help but be impressed with most of the true firs in native plant communities. Many of the true firs are found growing at higher elevations in the mountains of far western North America. The grace and symmetry of true firs adds much to the beauty of our mountain forest. True firs are not the easiest member of the pine family to culture as lawn or garden trees. These shade-tolerant trees find the bright sun and low relative humidity of the American Southwest a harsh environment for growth. For the gardener willing to learn about and provide a favorable environment for the growth of true firs, the trees will provide ample rewards. The cones of several of the species of true fir are spectacular displays of natural color before maturity. A fruiting Red Fir (*A. magnifica*) can resemble a candelabrum supporting red velvet candles. The Santa Lucia Fir (*A. venusta*) has a very restricted natural distribution in coastal California and is grown by native plant gardeners because it is rare and because of its handsome foliage and unusual cones. Do not expect cones on your true fir over night. Collect and grow these trees now so your grand-children can enjoy these ideal Christmas trees.

At maturity, fir cones are 3–10 inches long and typically cylindric to ovoid in shape. Each cone scale bears 2 seeds at its base. Ripening and seed dispersal occur the same year the seeds are set. The cone scales and seeds separate on the tree, leaving only the spikelike cone axis on the branch.

A mature fir seed has a large wing and is typically ovoid to oblong in shape. Most of the seed is occupied by a fleshy endosperm.

Most firs bear seeds at 20–30 years, with larger seed crops generally occurring at 2–4-year intervals. *Abies* seeds are relatively fragile and can be damaged easily. Seeds store best at low seed-moisture content. Fir seeds can be stored at low temperatures for 5 years. Under ordinary storage conditions, *Abies* seeds retain little or no viability after one year.

Dormancy of fir seeds appears to be due to both embryo dormancy and the seed covering (integument). There is considerable variation between seed lots in degree of dormancy. *Abies* seed is typically stratified under cool-moist conditions prior to germination. Specific pregermination treatment recommendations generally call for moist stratification at 34°–41°F for 14–28 days. Prior to large-scale sowing in nursery beds, presoaked seeds may be placed in polyethylene bags and refrigerated for the specified stratification.period. In the field, *Abies* seeds often germinate in and on melting snow banks.

Cedrus—True Cedars

The true cedars are not native to North America, but are widely planted in the relatively moderate climates of the United States. The true cedars provide rapid growing trees with handsome and unusual foliage. In locations where these trees are adapted they make spectacular lawn trees. Allow room for expansion for the true cedars become large trees.

The seeds and scales of true cedar cones fall apart on the tree. like those of true firs. The seeds are oily and do not keep well under ordinary storage conditions.

The seeds exhibit little or no dormancy and will germinate without pretreatment. However, prechilling or cold stratification at 37°–41°F for 14 days has been recommended.

Calocedrus—Incense Cedar

Incense Cedar (*C. decurrens*) is the only member of this Pacific Basin group of conifers that occurs in the United States. It occurs on the Pacific Slope from Mount Hood to Mexico. Incense Cedar is rapid growing and adapted to fairly harsh environmental conditions. We have an Incense Cedar in our yard that has persisted in an area of shallow soil. We find that Incense Cedar foliage is an attractive and aromatic addition to Christmas decorations.

Cones, each containing up to 4 seeds, hang singly, are scattered throughout the crown, and mature in one season. The winged seeds are about an inch long and nearly ⅓ as wide.

Stratification of seeds for 30–60 days at 37°–41°F has been found to increase the amount and rate of germination.

Chamaecyparis—White Cedar-Port Orford Cedar

Of the 6 species of *Chamaecyparis,* 3 are native to North America, 2 to Japan, and one to Taiwan. Because of their beauty and variety of form, White Cedars are often used for ornamental plantings, hedges, and windbreaks.

Cones mature in September and October at the end of the first growing season. Cones have from 6–12 scales, each scale bearing from 1–5 seeds with thin marginal wings.

Germination of *Chamaecyparis* seed is characteristically low, due in part to poor seed quality and also to various degrees of embryo dormancy. Satisfactory stratification techniques have not been developed for these seeds. A promising pretreatment appears to be warm-moist stratification for 30 days at an alternating temperature of 68°–86°F followed by 30 days' moist stratification at 40°F.

Cupressus—Cypress

The true cypresses are evergreen trees or shrubby trees native to warm-temperate areas of the Northern Hemisphere. Cypresses are commercially propagated mainly for landscaping. Some species of cypress such as the Macnab Cypress (*C. macnabiana*) can be found growing on harsh sites with soils derived from serpentine. Species of cypress may be the answer for native plant gardeners with problem soils.

Most cypresses have seratinous cones that can hang on the trees for many years without opening. Cones should be collected by cutting them from the limbs. Cones can be opened by boiling for 30–60 seconds.

Seeds should be cool-moist stratified for 30 days at 34°F. Stratified cypress seeds will readily germinate at 72°F.

Juniperus—Juniper

The junipers include about 50–70 species of evergreen trees or shrubs, widely distributed throughout the temperate and subtropical regions, mainly in the Northern Hemisphere. There is a lot of variability in growth form and scale color in natural populations of junipers. Several species of juniper are adapted to the semi-arid conditions of the Great Basin and Southwestern portions of the United States. Native plant gardeners have an opportunity to exploit these variable forms for gardens if you can germinate the seeds. The Western Juniper (*J. occidentalis*) when loaded with berries has the appearance of a luxuriant blue-gray wax candle.

Juniper berries are usually collected in the fall by stripping. Seed can be extracted by running the fruits through a macerator. For some species with resinous fruits, soaking in a weak lye solution for 1–2 days helps to separate the oily resinous pulp from the seeds.

Juniper seeds should be stored dry in sealed containers at 20°–40°F. A moisture content of 10–12% is considered satisfactory.

Germination is delayed in seeds of most junipers because of embryo dormancy and also in some cases by an impermeable seed coat or inhibitors in the seed coat and fruit. The most common preseeding treatment is to stratify at 41°F for 30–120 days.

Seed germination of most junipers is so poor that it is difficult to obtain any plants. A simple but effective procedure to obtain some seedlings involves seeding a relatively large quantity of seeds in a sandy medium in pots and placing the pots outdoors in the fall. In late spring the pots can be moved into the glass house, taking advantage of natural stratification.

Larix—Larch

The larches include 10 species of cone-bearing, deciduous trees widely distributed over the cooler regions of the Northern Hemisphere. Larches are not easy to grow outside of their range of natural distribution. For those lucky enough to live in the Far Northwest, Northeast, or Lake States, the culturing of larches is a most worthwhile gardening endeavor.

A ripe larch cone is made up of brownish woody scales, each of which bears 2 seeds at the base. The seeds are chiefly wind-dispersed, and the empty cones remain on the trees for an indefinite period. Larch seed is winged and nearly triangular in shape.

Larch cones should be collected in the fall as soon as they are ripe. They may either be picked by hand from standing trees or logging slash, or obtained from squirrel caches.

Seeds of most larch species germinate fairly well without pretreatment. If seeded in the spring, larch seeds are often cool-moist stratified 32°–41°F for 20–60 days. Seeds of Subalpine Larch (*Larix lyalli*) are often pretreated by soaking in 3% hydrogen peroxide for 24 hours.

Picea—Spruce

The spruces are medium to tall evergreen conifers. Members of the genus *Picea* occur throughout the North Temperate regions of the world. Spruces are very adapted for growth in gardens, even outside their natural distribution. There are many ornamental forms of our native spruce species that are widely used in landscaping. The rare Weeping Spruce (*P. breweriana*) is a challenge for the most adventurous of native plant gardeners. There are Weeping Spruce trees growing naturally where the High Road crosses Sugar Creek in Siskiyou County, California, but the timing of fruiting and cone maturity is up to the individual gardener. We have several of the species of spruce that are native to the northeastern United States and Canada growing in our garden at Reno, Nevada. They not only grow and survive in the wind and low humidity of Reno, they also have produced abundant cone crops as relatively juvenile trees.

The persistent cone scales may be rounded, pointed, notched, or reflexed at the ends. The pendant mature cones open on ripening to shed seeds during the autumn and winter. Cones remain on the trees for about one year, and some seeds may fall all year.

Seeds of most species of spruce germinate promptly without pretreatment, but cold stratification has been used for a few species. When seeds of some species are chilled under moist conditions, light is not required for germination. Conversely, exposure of imbibed seeds to light during germination may overcome dormancy in many seed lots without previous stratification.

Pinus—Pine

The genus *Pinus*, one of the largest and most important of the coniferous genera, comprises about 95 species and numerous varieties and hybrids. Pines are widely distributed, mostly in the Northern Hemisphere from sea level to timberline. About 41 species of pines are native to the United States. The native plant gardener can find a pine adapted to virtually every type of environment where conifers grow. There are common to extremely rare pines. Pines can be found growing naturally from seashore dunes to alpine parklands. Various geographic races or subspecies of the Lodgepole Pine (*P. murrayana*) complex grow from ocean-level to the alpine of western North America. Within populations of trees such as Ponderosa Pine (*P. ponderosa*) that characterize vast landscapes native plant gardeners can try and select for morphologic or adaptative characteristics in which they may have interest. Connoisseurs of pines have a single-needle (*P. monophylla*), 2-needle (*P. edulis*, among others), 3-needle (*P. jeffreyi*, among many), 4-needle (*P. quadrifolia*), and 5-needle (*P. lambertiana*, among many) in their gardens.

Cones and seeds of most species mature rapidly during late summer and fall of the second year. Cones of a few species mature during late winter of the second year or early spring of the third year. The interval between large cone crops is variable and depends on the species and environmental factors.

Most pines of temperate climates shed their seeds in the fall, and the seeds germinate promptly during the first spring. Pine seeds display highly variable germination behavior. The type and degree of dormancy vary among species and lots from the same seed source. Seeds of many species of pine ordinarily germinate satisfactorily without pre-treatment, but germination is greatly improved and hastened by first subjecting the seeds to cold stratification, especially if the seeds have been stored.

Stratification of pine seeds is accomplished by first soaking the seeds in water for 1–2 days and then placing them in a moist medium or in a plastic bag and holding them at a temperature between 30°–41°F. Seeds of some species may exhibit extreme dormancy that requires more than 60 days of stratification. The dormancy can be more complex; an anatomically immature embryo with a physiological block may be coupled with an impermeable seed coat. For seeds of these species, prolonged cold stratification for 6–9 months is required.

Pseudotsuga—Douglas Fir

P. menziesii, the major commercial species in North America, includes two geographic varieties. Coastal Douglas Fir (var. *menziesii*) is fast-growing, long-lived, and sometimes becomes over 300 feet tall and attains a diameter of 8–10 feet. Rocky Mountain Douglas Fir (var. *glauca*) is slender-growing, shorter-lived, and seldom exceeds 130 feet in height. The various forms of Douglas Fir are adapted to a large portion of western North America. The richly green foliage arranged in pyramidal crowns makes Douglas Fir a striking lawn tree. You can select for a variety of growth forms within this species.

There is a second species of *Pseudotsuga,* Big Cone Spruce (*P. macrocarpa*), that is native to the coastal mountains of southern California. Collecting and growing Big Cone Spruce trees from seeds is a challenge for the advanced native plant gardener.

The mature, pendant cones of *Pseudotsuga* are readily identified by their three-lobed bracts which protrude beyond the cone scales. Under each scale are borne two seeds which have relatively large wings.

Stratification (moist-cool) of Douglas Fir seeds often strikingly improves the rate of germination.

Sequoia sempervirens—Redwood

Redwood is one of the largest of the forest trees. Its natural range is in the summer fog belt of the Coast Range from Little Redwood Creek on the Chetco River in southwestern Oregon to Salmon Creek in the Santa Lucia Mountains of southern Monterey County, California. The Redwood is a coastal species largely confined in nature to the fog belt along the northern California coast. Despite their natural distribution Redwoods grow surprisingly well in the Central Valleys of California when they receive artificial irrigation. There is a grove of Redwoods in Bidwell Park at Chico, California, that has persisted since a University of California planting trial established early in this century. Redwood seedlings have established naturally beneath the trees in this grove.

Cones are ripe in September of the first year. The cones are small, about ½–1½ inches long. Seeds are mature when cone color changes from green to greenish yellow.

Redwood seeds require no pretreatment to induce germination. Germination capacity (total germination) is characteristically low because of the high percentage of empty seeds.

Sequoiadendron giganteum—Big Tree

This species grows to heights exceeding 250 feet in central California on the western slope of the Sierra Nevada in more or less isolated groves at 4,500–7,500 feet elevation. The Big Tree is one of the most successfully planted conifers in windy, dry, and. cold Reno, Nevada. The awl-shaped leaves resist the desiccation of winter cold winds. Even as young trees, the deeply furrowed bark and swollen base of Big Tree trunks suggest the potential greatness of these trees. Hold a tiny Big Tree seed in your hand and contemplate growing a tree 50 or more feet high in your lifetime. In the harsh climate of the Great Basin, this is hardly possible among conifers except for the Big Tree.

Cones may remain attached to the tree for several years, and much of the seed will be retained. The cones are elongated globes about 2–3½ inches long. As soon as the cones are removed from the tree, they start to open.

The old persistent cones can be collected at any time, but fresh cone collections should be made in August and later. Squirrels cut and cache cones that furnish a considerable quantity of seeds for collection.

Germination of Big Tree seed has ranged from 30–40%. Optimum constant temperatures for germination range from 60°–70°F.

Taxodium distichum—Bald Cypress

Bald Cypress is a large deciduous tree that often reaches a height of 130 feet at maturity. It is a handsome ornamental tree in climates with limited frost.

The globose cones turn from green to brownish purple as they mature in October to December. The cones are ½–1¼ inches in diameter and consist of a few 4-sided scales that break away irregularly after maturity. Each scale bears 2 irregularly shaped seeds that have thick, horny, warty coats and projecting flanges. Cones contain 18–30 seeds.

Bald Cypress seeds exhibit an apparent internal dormancy that can be overcome by 90 days of cold stratification, preceded by a 5-minute soak in ethyl alcohol. Soaking the seeds in water at 38°F for 90 days has also been successful.

Taxus—Yew

About 8 species of yew occur in the North Temperate Zone. They are non-resinous evergreen trees or shrubs useful chiefly for ornamental purposes.

The fruit of yews, which ripens in late summer or autumn, consists of a scarlet fleshy, cuplike aril bearing a single, hard seed. The seed has a large, very oily, white endosperm and a minute embryo.

To prevent losses to birds, yew fruits should be picked from the branches by hand as soon as they are ripe. Seeds are extracted by macerating the fleshy arils in water and floating off the pulp and empty seeds.

Yew seeds are slow to germinate; natural germination usually does not take place until the second year. Natural germination of many species may occur after the seeds pass through birds. Yew seeds have a strong but variable dormancy that can be broken by warm plus cold stratification. One recommendation is to hold the seeds for 90–210 days at 60°F followed by 60–120 days at 36°–41°F.

Thuja—Arborvitae

The genus *Thuja* includes 2 species native to North America and 4 Asian species. All are aromatic evergreen trees, but some also have shrubby forms. Northern White Cedar (*T. occidentalis*) and Western Red Cedar (*T. plicata*) are important timber species in the United States.

Cones may be picked by hand from standing or recently felled trees, or cones may be flailed or stripped onto a sheet of canvas, burlap, or plastic. A good time for collection is when seeds have become firm and most cones have turned from yellow to brown.

Seeds of the *Thuja* species generally germinate without special pretreatment. Dormant seed lots have been encountered occasionally; stratification in a moist medium at 34°–40°F for 30–60 days stimulated their prompt germination.

Torreya—Torreya

The genus *Torreya* includes 6 closely related species of trees found in North America, central and northern China, and Japan. These species are isolated relics of a pre-Pleistocene circumpolar flora. California Torreya (*T. california*) is found in 24 small areas scattered along the Coast Range and wet slope of the Sierra Nevada in California. The ovule of the California Torreya develops in a sessile fleshy aril-like structure. By the end of the second growing season the fertilized ovule forms a thinly fleshed green-to-purple drupelike fruit 1–1¾ inches long. California Torreya seeds require a long afterripening period before germination. Even after 3 months of stratification it required 6 to 8 months after sowings for seedlings to emerge. California Torreya is an excellent species for lovers of conifers to attempt to grow as a specimen tree. There is a tree in the lower Feather River Canyon of California (Highway U.S. 70) that is close to the highway. You will still need local instructions to find the tree.

Tsuga—Hemlock

The hemlocks are tall, straight evergreens with conical crowns and slender, horizontal-to-pendulous branches. Fourteen species have been reported, 4 occurring in the United States and the others in the Himalayas, China, Taiwan, and Japan. The hemlocks are graceful ornamental trees in the Pacific Northwest, but are difficult to culture in the warmer climates of the Southwest. A fitting challenge to the most advanced native plant gardener is the growing of a specimen tree from collected seed of Mountain Hemlock (*T. mertensiana*). In natural plant communities in the high Sierra Nevada the Mountain Hemlock is one of the most delicate and striking of the Western conifers. The slender, pendulous branches of Mountain Hemlock hang like soft lace when viewed against the stark alpine scenery of the high mountains.

Seed production of hemlock usually begins when trees are 20–30 years of age. Cones, which mature in one season, are small, pendant, globose to ovoid, with scales longer than the bracts. Seeds are nearly surrounded by their wings.

Dormancy is variable in *Tsuga*, with some seed lots requiring pre-germination treatment and others germinating satisfactorily without treatment. Cold-moist stratification clearly accelerates and improves total germination.

SUGGESTED ADDITIONAL READING

Conifers

Abies—Fir Trees
Germination: Franklin, J. F. 1974. *Abies*, pp. 168–182. Handbook 450. *Seeds of woody plants in the United States.* Forest Service, U.S. Dept. of Agric., Washington, D.C. (Hereafter cited as Handbook 450.)

Cedrus—True Cedars
Germination: Rudolf, P. O. 1974. Cedrus, pp. 291–294. Handbook 450.

Chamaecyparis—White Cedar
Germination: Harris, A. S. 1974. Chamaecyparis, pp. 316–319. Handbook 450.

Cupressus—Cypress
Germination: Johnson, L. C. 1974. Cupressus, pp. 363–369. Handbook 450.

Nursery Practices: Wolf, C. B. and W. W. Wagener. 1948. The New World Cypresses. Vol. I. Rancho Santa Ana Botanic Garden. El Aliso, Ca. 444 pp.

Juniperus—Juniper
Germination: Johnsen, T. N., Jr. and R. A. Alexander. 1974. Junipers, pp. 460–469. Handbook 450.

Larix—Larch
Germination: Rudolf, P. O. 1974. Larix, pp. 478–485. Handbook 450.

Libocedrus (Calocedrus)—Incense Cedar
Germination: Stein, W. I. 1974. Libocedrus decurrens Torr., pp. 494–499. Handbook 450.

Picea—Spruce
Germination: Safford, L. O. 1974. Picea, pp. 587–597. Handbook 450.

Nursery Practices: Heit, C. E. 1968. Propagation from seed. Part 18. Some western and exotic spruce species. American Nurseryman 127(8):12, 13, 63.

Pinus—Pine
Germination: Krugman, S. L. and J. L. Jenkinson. 1974. Pinus, pp. 598–638. Handbook 450.

Nursery Practices: Heit, C. E. Propagation from seed. Part 12. Growing choice, less common pines. American Nurseryman 127(2):14–15, 112–120.

dem Ouden, P. and B. K. Boom. 1965. Manual of cultivated conifers. Martinus Nijhoff. The Hague. 526 pp.

Pseudotsuga—Douglas Fir
Germination: Ouston, P. W. and W. I. Stein. 1974. Pseudotsuga, pp. 674–683. Handbook 450.

Nursery Practices: Stein, W. I. 1969. Summary of Douglas-fir nursery practices at 18 western nurseries in 1969. Pacific Northwest Forest and Range Expt. Sta., Forest Service, U.S. Dept. Agric.

Sequoia—Redwood
Germination: Boe, K. N. 1974. Sequoia sempervirens (D. Don)
                    Endl., pp. 764–766. Handbook 450.

Sequoiadendron—Big Tree
Germination: Boe, K. N. 1974. Sequoiadendron giganteaum
                    (Lindl.) Buch., pp. 767–768. Handbook 450.

Taxodium—Bald Cypress
Germination: Bonner, F. T. 1974. Taxodium distichum (L.) Rich.,
                    pp. 796–798. Handbook 450.

Taxus—Yew
Germination: Rudolf, P. O. 1974. Taxus, pp. 799–802. Handbook
                    450.

Thuja—Arborvitae
Germination: Schopmeyer, C. S. 1974. Thuja, pp. 805–809.
                    Handbook 450.

Nursery Practices: Stoeckeler, J. H. and G. W. Jones. 1957. Forest
                    nursery practices in the Lake States. Handbook
                    110. Forest Service, U.S. Dept. Agric., Washington,
                    D.C. 124 pp.

Torreya—Torreya
Germination: Roy, D. R. 1974. Torreya, pp. 815–816. Handbook
                    450.

Tsuga—Hemlock
Germination: Ruth, R. H. 1974. Tsuga, pp. 819–827. Handbook
                    450.

BROADLEAF TREES

Acer—Maple
    Maples are deciduous (rarely evergreen) trees comprising
approximately 148 species in North America, Asia, Europe, and
northern Africa. Maples are only rarely used in reforestation planting
for the production of timber, but Sugar Maple (*A. saccharum*) has been
planted for the production of syrup and sugar.
    Most maples produce seeds at an early age and bear almost
annually. The winged fruit of maples is called a samara. For many of the
maples that are used for lawn trees, samaras can be gathered from lawns,
parking lots, or even from the surface of pools or streams.
    Maple seeds are usually not extracted from the samaras. For
maximum germination, seeds of most maple species require a warm,
moist period of stratification for a few months prior to a cold, prechilling
period, but there is wide variation in pregermination treatments. Seed

lots of maples vary considerably in the time required for afterripening to be satisfied. Stratification under snow has been used successfully in Europe.

Aesculus—Buckeye-Horse Chestnut

The buckeyes include about 25 species of deciduous trees or shrubs. They are cultivated for their dense shade or ornamental flowers. They also provide wildlife habitat.

The fruit is a somewhat spiny or smooth, leathery, round or pear-shaped capsule with 3 cells. Each cell may bear a single seed. The ripe seeds are dark chocolate to chestnut brown in color, smooth and shining, and have a large, light-colored hilum resembling the pupil of an eye. Seeds of most buckeye species require cool-moist stratification to induce germination. Stratified buckeye seeds can be germinated at diurnally alternating temperatures of 86° and 68°F.

Alnus—Alder

This genera includes about 30 species of deciduous trees and shrubs. Their most common native habitats are high mountains, swamps, and bottomlands along streams. The alders are seldom cultivated, but offer a promising source of plant material for native plant gardeners. If you have a riparian habitat to vegetate, the alders should be considered. Many species of alder fix nitrogen symbiotically which may explain their ability to colonize harsh environments. Despite being rich in protein the browse of many species of alder is not preferred by deer. This can be an advantage if your garden is visited by large numbers of browsers. The catkins of the alders may attractive additions to dried arrangements.

Clusters of male and female catkins occur on the same tree in late winter or spring. Seeds are small nuts borne in pairs on the bracts of the strobiles. When released, the seeds are dispersed by wind or water.

The seeds of most alders germinate when they are fresh, without pretreatment. If the seeds of some alders are dried to a moisture content of 8–9 percent they become dormant. This dormancy can be broken by cool-moist stratification at 41°F for 180 days.

Arbutus menzieii—Pacific Madrone

Pacific Madrone is native to the Pacific Coast from British Columbia to California. This species has been occasionally planted as an ornamental.

The fruit of the madrone is a berry, ⅓–½ inch in diameter, bright red or orange red when ripe. The fleshy layer should be removed from either fresh or dried fruits before attempting germination. Seeds must be stratified in a moist medium at 33°–40°F to stimulate germination. Stratification for 60 days is usually required.

Betula—Birch

The birches consist of about 40 species of deciduous trees and shrubs occurring in the cooler parts of the Northern Hemisphere. Several species produce valuable lumber; others, because of their graceful growth habit, have handsome foliage and bark that are useful for ornamental plantings.

The flowers are monoecious (male and female flowers borne on the same tree) catkins. When they ripen in late summer or autumn, the fruits become brown and woody. Each scale may bear a single small, winged nut.

Birch seed is collected by picking the catkins while they are still green enough to hold together. They shatter readily and usually are put directly into bags.

Stratification of birch seeds is not necessary if the seeds are germinated with light. The light requirement can be overcome with cool-moist stratification.

Carya—Hickory

The *Carya* species are deciduous trees that are valuable for timber and food for wildlife.

Hickories are monoecious and flower in the spring. The fruits are ovoid, globose, or pear-shaped nuts enclosed in husk developed from the flower base.

Nuts can be collected from the ground after natural seed fall or after shaking the trees. Hickories exhibit embryo dormancy, which can be overcome by stratification in a moist medium at 33°–40°F for 30–150 days. Naked stratification in plastic bags is suitable for most species. Older seeds require less stratification.

Castanea—Chestnut

Chestnuts form a small genus of small to medium-sized deciduous trees. The American Chestnut (*C. dentata*) was formerly one of the most valuable timber species in the Appalachian region. The nuts were an important wildlife food. The chestnut blight has spread throughout the native range of the American Chestnut and eliminated it as a commercial species.

The fruiting structure of a chestnut is a spiny, globose burr that encloses 1–3 nuts.

Castanea seeds require a period of afterripening and cool-moist stratification to overcome dormancy. If seeds are planted in the fall, no pre-stratification is required.

Catalpa—Catalpa

The catalpas include about 10 species of deciduous or rarely evergreen trees. The Southern Catalpa (*C. bignoniodes*) and the Northern or Hardy Catalpa (*C. speciosa*) have been widely planted as lawn and shade trees.

Mature fruits are round, brown, 2-celled capsules, 6–12 inches long. In late winter or early spring, the capsules split into halves to disperse the seeds. Each capsule contains numerous oblong, thin, papery, winged seeds 1–2 inches long and about ¼ inch wide.

Seeds of catalpa germinate promptly without pretreatment. Total germination is often over 90% and occurs within 2 weeks.

Cornus—Dogwood

About 40 species of dogwood are native to the temperate regions of the Northern Hemisphere, and one is found in Peru. Most species are deciduous trees or shrubs useful chiefly for their ornamental qualities. The dogwoods constitute some of the most striking flowered trees native to North America. The Pacific Dogwood (*C. nutalli*) is widely distributed in the mountains of the Far West. This species and the Eastern species, Flowering Dogwood (*C. florida*) are often cultivated as ornamentals. In Medford, Oregon, the various forms of dogwood are grown as lawn trees, where they produce spectacular displays of flowers beneath an overhead canopy of native oaks. Unfortunately, the dogwoods are not adapted to the warmer, lower-elevation portions of the Southwest.

Fruits are globular to ovoid drupes ⅛–¼ inch in diameter, with a succulent or mealy flesh containing a single 2-celled and usually 2-seeded stone. In many stones only one seed is fully developed.

Dogwood stones can be sown without extracting them from the fruit. After collection, the fruit may be sown immediately or stratified for spring planting.

Corylus—Hazel

The hazels include about 16 species of large deciduous shrubs or small trees. The fruits are round, hard-shelled, brown or dark tan nuts. The nut is enclosed in a husk. Hazel seeds require 2 to 6 months of prechilling before germination will occur. In nurseries this can be accomplished by fall sowing.

Cotinus—Smoketree

American Smoketree (*C. obovatus*) is a small tree native to the southeastern United States. It is cultivated as an ornamental species.

The fruit is a dry, compressed drupe, light red-brown in color, containing a thick, bony stone.

Cotinus seeds have both an impermeable seed coat and an internal dormancy. They can be best stimulated to germinate by a combination of sulfuric acid treatment and cold stratification.

Crataegus—Hawthorn

Hawthorns in North America consist of perhaps 100–200 species of small trees and shrubs. The hawthorns are widely grown both for their ornamental flowers and fruits. The fruit of the hawthorn is a pome containing from 1–5 nutlets. Fruit of many species remains on the tree over winter.

Seeds of all *Crataegus* species have embryo dormancy and require treatment in a moist medium at low temperatures before germination will occur. The stratification period is quite long, ranging from 120–180 days.

Diospyrus—Persimmon

The Common Persimmon (*D. virginiana*) is a small to medium-sized deciduous tree, normally attaining a height of 30 to 60 feet at maturity.

The persimmon fruit is a plum-like berry with a persistent calyx and contains three to eight seeds. The fruits ripen during the fall, and the seeds disperse from the fruit during the winter.

Natural germination occurs in the spring but may be delayed 2 or 3 years. Dormancy can be broken by stratification at 37°F for 60–90 days. The germination is delayed by the seed covering that inhibits emergence of the radicle. This also limits water imbibition. Clipping the caps can result in high germination.

Seeds of Texas Persimmon (*D. texana*) germinate at warm incubation temperatures without pretreatment. No dormancy was observed and viability was not reduced by 2 years' storage. Seeds were not dependent on soil coverage for seedling establishment.

Elaeagnus—Russian Olive

Russian Olive (*E. angustifolia*) is widely planted for ornamental purposes and for wildlife habitat. The Russian Olive is one tough tree. It has become naturalized in the depths of the Great Basin both in Utah and Nevada. When planted in favorable environments the Russian Olive can be a pest, but when you need a rapidly growing tree in saline/alkaline situations the Russian Olive is hard to beat.

The fruit of the Russian Olive is a dry and indehiscent achene enveloped by persistent fleshy perianth. Seeds are often distributed by birds following consumption of the ripe fruits.

Seeds of Russian Olive should be scarified in sulfuric acid for ½ to one hour before being stratified at 34°F for 90 days.

Seeds of Silverberry (*Elaeagnus commutata*) have a dual dormancy mechanism. A germination inhibitor is present in the seed coat, and the seeds will not germinate in the presence of light (negative photoblastic). The seed coat germination-inhibiting substance can be readily leached by washing the seeds in warm water for 24 hours.

Eucalyptus—Eucalyptus

The genus *Eucalyptus* comprises more than 523 species. Eucalyptus are mainly native to Australia, but they are widely planted as ornamentals in temperate to warm climates.

The fruit of the *Eucalyptus* is a hemispherical, conical, oblong, or ovoid, hard woody capsule. The seeds are numerous and extremely small in most species.

Most eucalyptus seeds need no pretreatment to insure adequate germination if fresh seeds are used. A few species are dormant at collection and require cool-moist stratification.

Fagus—Beech

The beeches include 10 species of medium-sized deciduous trees native to the temperate regions of the Northern Hemisphere. They are valuable for their ornamental qualities and nuts, which are eaten by both men and wildlife.

The beech fruit consists of two or three one-seeded nuts surrounded by a prickly burr or husk developed from the floral involucre. Beech nuts may be shaken from the trees after frost has opened the burrs, or the nuts may be raked from the ground after they have fallen, usually from mid-September to November.

Beech seeds require cold stratification for prompt germination. Stratification at 37°–41°F for 40–50 days should be satisfactory.

Fraxinus—Ash

The ashes comprise a large genus of deciduous trees useful for forest products, wildlife habitat, and ornamentals. There are 9 species of ash native to North America.

Ash fruits are elongated, winged, single-seeded samaras that are borne in clusters. Fruits mature by late summer or fall and are wind-dispersed.

The seeds of most species of ash exhibit dormancy that is apparently due to both internal factors and to seed coat effects. Older seeds appear more dormant than freshly collected ones. The most successful germination enhancement techniques involve combinations of warm and cold stratification. The warm period consists of stratification at a diurnal temperature regime of 68° and 86°F daily for 30–90 days. The following cold stratification consists of 41°F for 60–90 days.

Gleditsia—Honey Locust

There are 12 species of *Gleditsia*, with 2 species native to North America. All are deciduous trees that are useful for timber production and wildlife food.

Honey Locust are members of the legume family, so their fruit is a pod. The pods ripen in the fall but often persist until winter. The seeds are small, flat, and brownish.

The hard seeds of Honey Locust must be treated to make them permeable before germination can occur. Soaking the seeds in either concentrated sulfuric acid or hot water has been used, but acid scarification is most effective. Soaking time in acid must be determined for each seed lot. The optimum time varies from 1–2 hours.

Gymnocladus dioicus—Kentucky Coffeetree
Kentucky Coffeetree is another member of the legume family. This medium-to-large deciduous tree is native to the northeastern United States.

Kentucky Coffeetree seeds are naturally hard, with impermeable seed coats. Acid treatment for these seeds consists of soaking the seeds in water at room temperature for 24 hours and then soaking in concentrated sulfuric acid for 2 hours. For small quantities of seed, filing through the outer seed coat with a hand file will give acceptable results.

A more recent study has shown that a 150-minute acid scarification provided earlier and greater germination than other scarification treatments. Soaking in water before scarification did not enhance germination.

Ilex—Holly
The hollies include about 300 species of deciduous or evergreen shrubs and trees. The hollies are grown commercially for their shiny, prickly foliage and brightly colored berries. A holly bush is a great attraction for seed-eating birds. The nearly white wood of hollies is prized for inlays in construction of fine furniture. Holly fruits are rounded, berrylike drupes that contain 2–9 bony, one-seeded, flattened nutlets. The nutlets are extracted by running the fruits through a macerator with water.

Ilex seeds exhibit a deep dormancy that is caused partly by the hard endocarp surrounding the seed coat and partly by conditions in the embryo. The seeds of American Holly (*I. opaca*) contain an immature embryo that requires 16 months–3 years of afterripening. Some benefit may be obtained from warm-moist stratification for 60 days (68° and 86°F diurnally) followed by 60 days' cool-moist stratification at 41°F.

Juglans—Walnut
The walnut includes about 15 species of deciduous trees or large shrubs native to temperate areas in the Northern Hemisphere. The walnuts are of multiple value to man. The wood of many species is prized for fine furniture. The nuts are equally prized. In the hot Central Valleys of California, the California Black Walnut (*J. hindsii*) lines the streets and shelters the older residential areas from the summer sun. It is doubtful that any truly native stands of California Black Walnut exist. Many "naturally occurring" stands are actually the sites of former Indian habitations. Drive the back streets of Chico, California, on a summer day if you want to experience the California Black Walnut as a grand shade

tree.

The fruit is a nut enclosed in an indehiscent, thick husk that develops from a floral involucre. The nut has a hard shell. The four-lobed seed remains within the shell during germination.

Naturally, germination usually occurs in the spring following fall seeding. Most *Juglans* species have a dormant embryo. Dormancy can be broken by stratification at temperatures of 34°–41°F for 30–190 days depending on the species.

Liquidambar styraciflua—Sweetgum

Sweetgum is native to a variety of sites from Connecticut to Nicaragua. Sweetgum is very valuable for pulp, lumber, and veneer. The tree has been widely planted as an ornamental.

The female flowers of sweetgum are borne in globose heads, which become the multiple heads of two-celled capsules. The beaklike capsules open in September, dispersing small, winged seeds.

Sweetgum seeds exhibit only a shallow dormancy, but germination rate is considerably increased by cold-moist stratification. Naked stratification in plastic bags at 33°F for 15–90 days has been effective.

Liriodendron tupilifera—Yellow Poplar

Yellow Poplar is native to the eastern United States. Yellow Poplar is very valuable for lumber and veneer. It is a good honey tree and is planted extensively as an ornamental.

The fruit is an elongated cone composed of closely overlapped carpels that are dry, woody, and winged.

Seeds to be sown in the spring and seeds taken from dry storage need pregermination treatments to overcome an internal dormancy. Several treatments have proven satisfactory: (a) storage in moist, well-drained pits or mounds of soil, sand, peat, or mixtures of these media from over winter to as long as 3 years; (b) cool-moist stratification in bags of peat moss or sand for 60–90 days; or (c) cool-moist stratification in plastic bags for 140–170 days. Recommended temperatures for stratification are 35°–36°F.

Lithocarpus densiflora—Tanoak

Tanoak is an evergreen hardwood native to California and Oregon. Tanoak bark was used to furnish the best tannage known for the production of heavy leather.

The fruit of Tanoak is an acorn that ripens in the second autumn. Acorns of Tanoak should be seeded in peat or sand soon after they are collected to avoid dormancy.

Maclura pomifera—Osage Orange

Osage Orange is native to southern Arkansas, southeastern Oklahoma, and eastern Texas. This tree has been widely planted and has become naturalized in the eastern United States. Osage Orange is occasionally planted in the Far West as a specimen tree. In the south central United States this species has been widely used for hedge plantings.

The yellow-green fruits of Osage Orange, which are 4–5 inches in diameter, are composed of one-seeded drupelets.

Fruits should be collected as soon as they fall. Seeds may be extracted by macerating the fruits and screening out the pulp.

Osage Orange seeds exhibit a slight dormancy that may be overcome by stratification for 30 days at 41°F.

Magnolia—Magnolia

The genus *Magnolia* is composed of about 35 species of deciduous or evergreen trees or shrubs which occur in North and South America, eastern Asia, and the Himalayas. Some of the species produce valuable timber. Many magnolias are used in ornamental planting because of their showy flowers and fruits, and attractive foliage.

The red to rusty brown cone-like fruits of magnolias consist of several coalescent one to two-seeded fleshy follicles which ripen in late summer to early fall.

Magnolia seeds exhibit embryo dormancy which can be overcome by 3–6 months of low temperature stratification (32°–41°F). Fall sowing produces natural stratification.

Malus—Apple

The apples consist of about 25 species of deciduous trees or shrubs native to the temperate regions of North America, Europe, and Asia. Commercial apple trees are propagated by budding or grafting. Interesting variations in flower color can be obtained by growing wild apples from seed.

The apple fruit is a fleshy pome in which 3–5 carpels (usually 5) are embedded. Each carpel contains 1 or 2 seeds. Apple seeds display dormancy which can be overcome by cold stratification. Stratification temperatures are 37°–41°F for 30–120 days.

Morus—Mulberry

The mulberries consist of about 12 species of deciduous trees and shrubs native to temperate and subtropical regions of Asia, Europe, and North America.

Flowers of mulberries appear as stalked catkins. The multiple fruits are composed of many small, closely appressed drupes.

Germination is highly variable, depending on seed lots. Some lots will germinate completely while others have seeds with impermeable seed coats and embryo dormancy. For fall sowing, a 100-hour presoak in water is beneficial. For spring sowing, stratification for 30–90 days at 30°–41°F is recommended.

Myrica—Bayberry

The genus *Myrica* is composed of evergreen shrubs and small trees up to 40 feet high. California Wax-Myrtle (*M. california*) is a large evergreen shrub that makes an excellent ornamental when grown in moist habitats. This species is found growing natively from the Santa Monica Mountains of southern California to Washington. Fruits are small, globose, dry drupes heavily coated with wax.

Germination of cleaned Bayberry seeds is accelerated and increased by stratification at 34°–40°F for 60–90 days.

Nyssa—Tupelo

Three deciduous arboreal species of *Nyssa* native to North America are useful for timber production, wildlife food, and honey production.

The fruits of *Nyssa* are thin-fleshed, oblong drupes. They ripen in the fall.

Nyssa seeds exhibit embryo dormancy, and they benefit from cold-moist stratification. Good germination has been reported after only 30 days' cool-moist stratification, but periods of up to 120 days may be needed.

Platanus—Sycamore

Sycamores are deciduous trees that range from 80–140 feet tall at maturity. American Sycamore (*P. occidentalis*) is one of the largest and most valuable timber species in the eastern United States. Sycamores are widely planted as ornamentals.

The fruit of sycamores is an elongated, chestnut-brown, single-seeded achene with a hairy tuft at the base.

Cool-moist stratification for 60–90 days at 40°F in sand, peat, or sandy loam has been reported in enhancing germination of California Sycamore (*P. racemosa*) seeds. American Sycamore and Oriental Planetree (*P. orientalis*) seeds do not require pregermination treatment.

Populus—Poplar

The genus *Populus* includes about 30 species of medium to large deciduous trees native to North America, Europe, North Africa, and Asia. Poplars are important pulpwood, lumber, and veneer species. Several species and cultivars of poplar are widely planted as ornamentals.

The fruit of the *Populus* species consist of 2–4-valved capsules arranged in catkins. Branches bearing near-mature catkins can be brought into a warm room or greenhouse and placed in water to allow the capsules to open. The seeds are covered with fine, silky hairs.

Germination of poplar seeds is rather tricky. Seeds may show viability by breaking their seed coats when placed in germination test and yet not have enough vitality to produce a normal plant. Fresh seeds usually begin to germinate within 12 hours after being placed on a saturated seedbed.

Prunus—Cherry, Peach, Plum

The genus Prunus is one of the most important genera of woody plants. Its 5 well-defined subgenera include: (a) the plums and apricots, (b) the umbellate cherries, (c) the deciduous racemose cherries, (d) the evergreen racemose or laurel cherries, and (e) the almonds and peaches.

The fruit of the *Prunus* species is a one-seeded drupe. The drupe is thick and fleshy, except in the almonds, and has a bony stone or pit.

Prunus fruits should be collected when fully mature. Generally, it is desirable to clean the seeds from the pulp.

Prunus seeds have embryo dormancy and require a period of after-ripening and stratification in the presence of moisture and oxygen to overcome the dormancy. Because of the stony endocarp, *Prunus* species are often considered to be hard-seeded, but this is not true. However, removal of the endocarp may hasten or increase germination in some species. *Prunus* seeds have been stratified in peat, sand-peat mixtures, or activated charcoal. Stratification temperatures range from 36°–41°F, with the duration depending on the species.

Ptelea trifoliata—Common Hoptree

Hoptree is a shrub or small tree up to 25 feet tall. It has some value for wildlife, shelterbelt, and environmental plantings. The natural range of hoptree. is from southern Ontario to eastern Kansas and northern Florida. Fruits of the hoptree are reddish-brown samaras. Seed germination is quite slow, probably because of embryo dormancy. Germination can be hastened by cool-moist stratification at 41°F for 3–4 months.

Quercus—Oak

The genus *Quercus,* with many species of deciduous and evergreen trees and shrubs, is the most important aggregation of hardwoods found on the North American continent, if not in the entire Northern Hemisphere. There are about 70 species native to the United States, and 58 of these reach tree size. The uses of oak include almost everything that mankind has ever derived from trees—timber, food for man and other wild and domestic animals, fuel, watershed protection, shade and beauty, tannin and extractives, and cork.

The fruit of oaks is an acorn or nut, which matures in one year (White Oaks) or two years (Black Oaks). Acorns are one-seeded or, rarely, two-seeded and occur singly or in clusters of 2–5.

Acorns may be collected from the ground, or flailed or shaken from branches. California Black Oak (*Q. kelloggii*) acorns require prompt collection because of a mold that often affects fallen acorns and destroys the cotyledons. Birds and small mammals consume large quantities of acorns. Fallen acorns, termed mast, are consumed by numerous large herbivores. The only extraction required of acorns before storage or sowing is removal of loose caps, twigs, and other debris.

With few exceptions, acorns of the white oak group have little or no dormancy and will germinate almost immediately after falling. Acorns of the black oak group exhibit embryo dormancy and germinate the following spring after fall sowing. Stratification is required before germination testing or spring sowing of acorns of the black oaks. For best results, stratification should be in moist, well-drained sand, or sand and peat for 30–90 days at temperatures of 30°–41°F. The acorns of several black oaks will begin germination at stratification temperatures if the duration exceeds 30–45 days. Shoot emergence will not occur at stratification temperatures.

Rhamnus—Buckthorn

The genus *Rhamnus* consists of about 100 species and many varieties. The genus is chiefly native to the temperate and warm regions of the Northern Hemisphere. Buckthorns have been used as ornamentals. The bark of Cascara Sagrada (*R. purshiana*) is used extensively in drug manufacturing.

The buckthorn fruit is a berry-like drupe containing 2–4 nutlike seeds.

Considerable variability seems to exist in the need for pregermination treatments of buckthorn seed, both between species and within some species. Fresh seeds of Alder Buckthorn (*R. alnifolius*), California Buckthorn (*R. california*), and Holly Leaf Buckthorn (*R. crocea* var. *ilicifolia*) apparently require no pregermination treatment. Stored seeds of these species and other species require cold stratification to break embryo dormancy. Cool-moist stratification requirements are temperatures of 34°–41°F for 15–110 days. Seeds of some species have hard seed coats requiring acid scarification. Scarification in concentrated H_2SO_4 for 20 minutes is suggested.

Robinia—Locust

The genus *Robinia* includes about 20 species which are native to the United States and Mexico. The locusts are members of the legume family. The locust fruit is a legume (pod) which ripens in the fall and contains 4–10 dark brown seeds. Locust are hardy, rapidly growing trees that were planted around many Western homesteads to soften the harsh, raw environment. Today you can find naturalized groves of locust marking long-abandoned farmsteads.

Dormancy in *Robinia* seeds is due to impermeable seed coats. Either mechanical or acid scarification or soaking in hot water (boiling or nearly boiling) is necessary. The duration of acid (H_2SO_4) scarification depends on the seed lot and must be determined by experimentation for each lot. The range in duration of acid scarification is 10–120 minutes.

Salix—Willow

The willows consist of about 300 species of deciduous trees and shrubs widely distributed in both hemispheres from the arctic region to South Africa and southern Chile. Of the some 70 North American species, about 30 attain tree size and form, and many are valuable for wood products. Willows are valuable browse species.

The willow fruit is a capsule occurring in elongated clusters. Each capsule contains many minute, hairy seeds. These usually ripen in early summer, but the seeds of some species mature in the fall.

Under natural conditions, willow seeds usually germinate in 12–24 hours on moist sand. Seed dormancy has not been observed in any species that disperses seeds in the spring. Some willows disperse their seeds in the late fall. These species usually grow about timber or beyond the tree line in the Arctic. When the seeds of these species are mature they are not dormant, but they acquire a dormancy hanging in the catkins. This dormancy is temperature-related. The seeds will not germinate at low temperatures. Cool-moist stratification breaks down this dormancy. Seeds must be sown immediately after collection.

Sapium sebiferum—Tallowtree

Tallowtree is a small deciduous tree which is planted as an ornamental in the Southeast and other environments where frost is rare. This native of China readily escapes from cultivation and is common along roadsides of the Gulf Coast. The fruit is a 3-lobed capsule containing 2 or 3 white, waxy seeds. Freshly collected seeds have moderate germination without pretreatment.

Sassafras albidum—Sassafras

Sassafras is a small to medium-sized tree native to the northeastern United States. Common Sassafras (S. varifolium) is often grown in parks for its ornamental flowers, fruit, and fall coloration of the leaves. The fruit is a small drupe dispersed by birds. The seeds exhibit strong embryo dormancy that can be overcome by cool-moist stratification for 120 days at 41°F.

Sorbus—Mountain Ash

The mountain ashes include more than 80 species of deciduous trees and shrubs distributed through the Northern Hemisphere. Their graceful foliage and showy, brightly-colored flowers make them especially sought for ornamental plantings. The mountain ash are important food plants for wildlife.

The showy orange-red to bright red fruits are berrylike pomes. Each cell of the fruit contains 1–2 small brown seeds. Sorbus seeds require 60 days or more of cool-moist stratification at 30°–41°F in moist sand or peat moss.

Tilia—Basswood or Linden

The lindens include about 30 species distributed in the Northern Hemisphere. Most are small to moderate-sized deciduous trees. They are valuable as timber trees for lumber and veneer. American Basswood (*T. americana*) has been widely planted as an ornamental.

The fruits of *Tilia* species are small, rounded, indehiscent capsules. Fruits of the American basswood can be mechanically broken to extract the seeds. Seeds of *Tilia* species show delayed germination because of impermeable seed coats, embryo dormancy, and a tough pericarp. Seed treatments that consistently result in good germination have not been developed.

Ulmus—Elm

About 20 species of elm are native to the Northern Hemisphere. None occur in western North America north of Mexico. Exotic species of elms have been widely planted as ornamentals and have become naturalized in many parts of the United States. The Dutch elm disease, caused by the fungus *Eratocystis ulmi,* and phloem necrosis, caused by the virus *Morus ulmi,* have killed many elms.

The fruit of elms, a samara, ripens a few weeks after flowering. The fruit consist of a compressed nutlet surrounded by a membranous wing. Dispersal is by wind.

Under natural conditions, elm seeds that ripen in the spring usually germinate the same growing season, but some seeds always remain dormant until the next season.

Umbellularia california—California Laurel

California Laurel is also known by many common names, including Oregon Myrtle. The tree is distributed from Central Oregon through California. The California Laurel wood is sold in Oregon as the highly figured Myrtlewood. This is a very attractive tree adapted to mid-elevation sites in California and much of western Oregon.

The small, pale yellow blossoms grow on short-stemmed umbels. The fruit is an oblong drupe that becomes dark purple when ripe.

California Laurel seeds appear to have an afterripening requirement that lasts up to three months after maturity. Fair germination can be obtained from fall-sown seeds on light-textured beds.

SUGGESTED ADDITIONAL READING

Broadleaf Trees

Acacia—Acacia
Germination: Whitesell, C. D. 1974. *Acacia,* pp. 184–186. Handbook 450.

Nursery Practices: Parry, M. S. 1956. *Treeplanting practices in tropical Africa.* FAO. Forage Development Paper 8. Rome. 302 pp.

Acer—Maple
Germination: Carl, C. M., Jr. 1983. Stratification of sugar maple seeds. *Tree Planters Notes* 34(1):25–26.

                           Olson, D. F., Jr. and W. J. Gabriel. 1974. *Acer,* pp. 187–194. Handbook 450.

Nursery Practices: Godman, R. M. 1965. Sugar maple (*Acer saccharum* Marsh.), pp. 66–73. Silvics of forest trees of the United States. Agric. Handbook 271. U.S. Dept. Agric.

Aesculus—Buckeye, Horse Chestnut
Germination: Rudolf, P. O. 1974. *Aesculus,* pp. 195–199. Handbook 450.

Alnus—Alder
Ecology: McDermott, R. E. 1959. Ecology of *Alnus glutinosa* (L.) Gaertn. Part VII. Establishment of alder by direct seeding of shallow blanket bog. *J. Ecology* 47:615–618.
Germination: Schopmeyer, C. S. 1974. *Alnus,* pp. 206–211. Handbook 450.

Arbutus—Pacific Madrone
Germination: Roy, D. F. 1974. *Arbutus.* Handbook 450.

Betula—Birch
Germination: Brinkman, K. A. *Betula,* pp. 252–257. Handbook 450.

Carya—Hickory
Germination: Bonner, F. T. and L. C. Maisenholder. 1974. *Carya,* pp. 269–272. Handbook 450.

Nursery Practices: Heit, C. E. 1942. Field germination and propagation of various hardwoods. New York State Conservation Dept. Notes on Forest Investigation No. 43.

Castanea—Chestnut

Germination: Sander, I. L. 1974. *Castanea,* pp. 273–275. Handbook 450.

Nursery Practices: Jaynes, R. A. and A. H. Graves. 1963. Connecticut hybrid chestnuts and their culture. Connecticut Agric. Expt. St. Bull. 657. New Haven, Conn. 29 pp.

Catalpa—Catalpa

Germination: Bonner, F. T. and D. L. Graney. 1974. *Catalpa,* pp. 281–283. Handbook 450.

Nursery Practices: Engstrom, H. E. and J. H. Stoeckeler. 1941. Nursery practice for trees and shrubs suitable for planting on the prairie-plains. Misc. Publ. 434. U.S. Dept. Agric. Wash. D.C. 159 pp.

Cornus—Dogwood

Germination: Brinkman, K. A. 1974. *Cornus,* pp. 336–342. Handbook 450.

Nursery Practices: Heit, C. E. 1968. Propagation from seed, par. 15. Fall planting of shrub seeds for successful seedling production. *American Nurseryman* 128(4):8–10, 70–80.

Corylus—Hazel, Filbert

Germination: Brinkman, K. A. 1974. *Corylus,* pp. 343–345. Handbook 450.

Cotinus—Smoketree

Germination: Rudolf, P. O. 1974. *Cotinus,* pp. 346–348. Handbook 450.

Crataegus—Hawthorn

Germination: Brinkman, K. A. 1974. *Crataegus,* pp. 356–360. Handbook 450.

Nursery Practices: Heit, C. E. 1967. Fall planting of fruit and hardwood seeds. *American Nurseryman* 126(4):12–13, 88–90.

Diospyros—Persimmon

Germination: Everitt, J. H. 1984. Germination of Texas persimmon seed. *J. Range Manage.* 37:189–191.

Olson, D. F., Jr. and R. L. Barnes. 1974. *Diospyros virginiana,* pp. 373–375. Handbook 450.

Elaeagnus—Russian Olive

Germination: Fung, M. Y. P. 1984. Silver seed pretreatment and germination techniques. *Tree Planters Notes* 35(3)32–33.

Olson, D. F., Jr. 1974. *Elaeagnus,* pp. 376–379. Handbook 450.

Eucalyptus—Eucalyptus
Germination: Krugman, S. L. 1974. Eucalyptus, pp. 384–392. Handbook 450.

Nursery Practices: Holmes, D. A. and A. G. Floyd. 1969. Nursery techniques for raising eucalyptus in jiffy pots on the New South Wales north coast. Commonwealth Forestry New South Wales Research Note 72. 15 pp.

Fagus—Beech
Germination: Rudolf, P. O. and W. B. Leak. 1974. *Fagus,* pp. 401–405. Handbook 450.

Nursery Practices: Heit, C. E. 1967. Propagation from seed. Part 8. Fall planting of fruit and hardwood seeds. *American Nurseryman* 126(4):12–13, 85–90.

Fraxinus—Ash
Germination: Bonner, F. T. 1974. *Fraxinus,* pp. 411–416. Handbook 450.

Nursery Practices: Eliason, E. J. 1965. Treatment of forest tree seed to overcome dormancy prior to direct seeding. Indirect seeding in the Northeast. H. Symposium. Univer. Mass. Exp. Stn. Bull. pp. 87–91.

Gram, W. H. 1984. Resewing treatments and storage for green ash seeds. *Tree Planters Notes* 35(1)20–21.

Tinus, R. W. 1982. Effects of dewinging soaking, stratification, and growth regulators on germination of green ash seed. *Canadian Journal of Forest Research* 12:931–935.

Gleditsia—Honey Locust
Germination: Bonner, F. T., J. D. Burton, and H. C. Grigsby. 1974. *Gleditsia,* pp. 431–433. Handbook 450.

Gymnocladus—Kentucky Coffeetree
Germination: Sander, I. L. 1974. *Gymnocladus dioicus* (L.) K. Koch, pp. 439–440. Handbook 450.

Yeiser, J. L. 1983. Germination pretreatments and seedcoat impermeability for the Kentucky coffeetree. *Tree Planters Notes* 34(2):33.

Ilex—Holly
Germination: Bonner, F. T. 1974. *Ilex,* pp. 450–453. Handbook
                      450.

Nursery Practices: Hartmann, H. T. and D. E. Kester. 1968. *Plant
                      propagation: principles and practices.* Ed. 2. Prentice-
                      Hall, Englewood Cliffs, N.J.

Juglans—Walnut
Germination: Brinkman, K. A. 1974. *Juglans,* pp. 454–459.
                      Handbook 450.

Nursery Practices: Brinkman, K. A. 1965. Black walnut (*Juglans nigra*
                      L.), pp. 203–207. Silvics of Forest Trees of the
                      United States. Handbook 271. U.S. Dept. Agric.
                      Clark, F. B. 1965. Butternut (*Juglans cinerea* L.), pp.
                      208–210. Silvics of Forest Trees of the United
                      States. Handbook 271. U.S. Dept. Agric.

Liquidambar—Sweetgum
Germination: Bonner, F. T. 1974. *Liquidambar straciflua* L., pp.
                      505–507. Handbook 450.

Nursery Practices: Vande, L. F. 1964. Nursery practices for southern
                      oaks and gums. *Tree Planters Notes* No. 65, pp. 24–
                      26.

Liriodendron—Yellow Poplar
Germination: Bonner, F. T. and T. E. Russell. 1974. *Liriodendron
                      tulipifera* L., pp. 508–511. Handbook 450.

Nursery Practices: Sluder, E. R. 1964. Quality of yellow poplar
                      planting stock varies by mother tree and seedbed
                      density. *Tree Planters Notes* No. 65, pp. 16–19.

Lithocarpus—Tanoak
Germination: Roy, D. F. 1974. *Lithocarpus densiflorus* (Hook &
                      Arnill) Redd., pp. 512–514. Handbook 450.

Maclura—Osage Orange
Germination: Bonner, F. T. and E. R. Ferguson. 1974. *Maclura
                      pomifera* (Raf.) Schneid., pp. 525–526. Handbook
                      450.

Nursery Practices: Engstrom, H. E. and J. H. Stoeckeler. 1941.
                      Nursery Practices for trees and shrubs suitable for
                      planting on the prairie-plains. Misc. Publ. 434. U.S.
                      Dept. Agric. Washington, D.C. p. 159.

Magnolia—Magnolia
Germination: Olson, D. F., R. L. Barnes, and L. Jones. 1974.
                      Magnolia, pp. 527–530. Handbook 450.

Nursery Practices:	Heit, C. E. 1939. Seed treatment and nursery practices with cucumber (*Magnolia accuminata*). New York State Conservation Dept. Notes on Forest Investigations No. 20, p. 2.

Malus—Apple

Germination:	Crossley, J. A. 1974. *Malus,* pp. 531–536. Handbook 450.

Morus—Mulberry

Germination:	Reed, R. A. and R. L. Barnes. 1974. *Morus,* pp. 544–547. Handbook 450.

Myrica—Bayberry

Germination:	Krochmal, A. 1974. *Myrica,* pp. 548–558. Handbook 450.

Nyssa—Tupelo

Germination:	Bonner, F. T. 1974. *Nyssa,* pp. 554–557. Handbook 450.

Plantanus—Sycamore

Germination:	Bonner, F. T. 1974. *Plantanus,* pp. 641–644. Handbook 450.
Nursery Practices:	Briscoe, C. B. 1969. Establishment and early care of sycamore plantations. Res. Paper SO–50. Forest Service, U.S. Dept. Agric.

Populus—Poplar

Germination:	Schreiner, E. J. 1974. *Populus,* pp. 645–655. Handbook 450.
Nursery Practices:	Fowells, H. A. (compiler). 1965. Silvics of forest trees of the United States. Handbook 271. Forest Service, U.S. Dept. Agric. 762 pp.
	Gammage, J. L. and L. C. Maisenhelder. 1962. Easy way to sow cottonwood nursery beds. *Tree Planters Notes* No. 51, pp. 19–20.

Prunus—Cherry, Peach, and Plum

Germination:	Grisez, T. J. 1974. *Prunus,* pp. 658–673. Handbook 450.
Nursery Practices:	Heit, C. H. 1945. Fall planting of hardwood tree seed. *Farm Research* (New York State Experiment Station) 11(3):14–15.

Ptelea—Common Hoptree

Germination:	Brinkman, K. A. and R. C. Schlesinger. 1974. *Ptelea trifoliata* L., pp. 684–685. Handbook 450.

Quercus—Oak
Germination: Olson, D. F., Jr. 1974. *Quercus,* pp. 642–703.
                      Handbook 450.

Rhamnus—Buckthorn
Germination: Hubbard, R. L. 1974. *Rhamnus,* pp. 704–708.
                      Handbook 450.

Robinia—Locust
Germination: Olson, D. F., Jr. 1974. *Robinia,* pp. 728–731.
                      Handbook 450.

Nursery Practices: Roberts, D. R. and S. B. Carpenter. 1983. The
                      influence of seed scarification and site preparation
                      on establishment of black locust on surface-mined
                      sites. *Tree Planters Notes* 34(3):28.

Salix—Willow
Germination: Brinkman, K. A. 1974. *Salix,* pp. 746–750.
                      Handbook 450.

Sorbus—Mountain Ash
Germination: Harris, A. S. and W. I. Stein. 1974. *Sorbus,* pp. 780–
                      784. Handbook 450.

Nursery Practices: Heit, C. E. 1967. Propagation from seed. Part 8. Fall
                      planting of fruit and hardwood seeds. *American
                      Nurseryman* 126(4):12–13, 85–90.

Tilia—Basswood, Linden
Germination: Brinkman, K. A. 1974. *Tilia,* pp. 810–812.
                      Handbook 450.

Ulmus—Elm
Germination: Brinkman, K. A. 1974. *Ulmus,* pp. 829–834.
                      Handbook 450.

Nursery Practices: Goor, H. Y. 1955. Treeplanting practices for arid
                      areas. Forest Development Paper 6. FAO. Rome.
                      126 pp.

Umbellularia—California Laurel
Germination: Stein, W. I. 1974. *Umbellularia,* pp. 835–839.
                      Handbook 450.

CHAPTER 10

Germination of Seeds of Shrubs

Several shrub species were covered in the second edition of the *Seeds of Woody Plants in the United States*. However, in comparison to trees species, virtually nothing is known about the collection, threshing, and germination of seeds of shrubs.

There are a lot of valuable shrubs that are important browse species and useful in environmental plantings. Many shrubs are used as ornamentals, but these species are usually propagated vegetatively.

On extensive areas of rangelands there are several species of shrubs that would be very desirable to establish by direct seeding. Unfortunately, no desirable browse species can be consistently established by direct seeding on rangelands.

Acacia—Acacia

Seeds of Twisted Acacia (*A. schaffneri*) required scarification in concentrated sulfuric acid for marked germination. Scarification for 45 minutes is required.

Soaking seeds of Blackbrush (*A. rigidula*) in concentrated sulfuric acid for 30 minutes enhanced germination. Light was not required for germination. Seeds of Guajillo (*A. berlandieri*) germinated without pretreatment.

Acamptopappus sphaerocephalus—Gray Goldenhead

Goldenhead is a low, much-branched shrub found on the margins of the Mojave Desert and the deserts of the Great Basin. This member of the sunflower family has achenes for fruits.

The achenes are difficult to collect because of the scattered nature of goldenhead stands. The achenes are very dormant with only limited germination at low incubation temperatures.

Adenostoma—Chamise

Adenostema is a genus of the rose family that consists of two species found in the chaparral vegetation of California. Chamise (*A. fasiculatum*) dominates vast areas of chaparral, where it constitutes an extreme fire hazard. Red Shank (*A. sparsifolium*) is another evergreen species found in chaparral communities.

The fruit is an achene enclosed in an indurate flower tube. The achenes have hard seed coats that require scarification in H_2SO_4 for 15 minutes. An alternative germination enhancement procedure involves seeding the seeds in soil in flats and burning pine needles on the soil surface.

Ailanthus altissima—Tree of Heaven

Native to China, Tree of Heaven can become a short to medium-tall deciduous tree. In the Gold Country of northern California, Tree of Heaven is naturalized on the old diggings and in vacant lots among the old buildings. The species forms large clumps from root stocks. In the fall, leaves turn brilliant red. The species is naturalized from California to Massachusetts and adjacent Canada. The Tree of Heaven is a questionable ornamental species because the foliage has a disagreeable odor, and it can rapidly become a weed.

The fruit is a one-celled spirally twisted samara. Fruits can be stripped from the plants or flailed onto drop cloths. Seeds require cold-moist stratification at 41°F for 60 days.

Amelanchier—Serviceberry

Many of the valuable browse species are members of the rose family. *Amelanchier* is an example. Not the most valuable of browse species, the serviceberries are sometimes suspect of being toxic because of their prussic acid content. The serviceberries include about 25 species of small deciduous trees and shrubs native to North America, Europe, and Asia.

The perfect, white flowers of serviceberry appear in terminal clusters early in the spring. Fruits are berrylike pomes that turn dark purple or black when they ripen. Each fruit contains from 4–10 small seeds, although some of these are usually abortive. Fertile seeds are dark brown with a leathery seed coat.

Fruits of serviceberry should be collected as soon as possible when they ripen to minimize losses to birds and small mammals. The fruits are picked or stripped from the branches. The seeds should be extracted at once to avoid overheating of the high-moisture-content fruits.

Seed extraction is usually accomplished by macerating the fruits in water and washing them over screens. This removes most of the pulp so the seeds can be safely dried. After drying and rubbing through screens, the remaining debris can be separated by running the material on an air-screen.

The seeds of all serviceberry species have embryo dormancy that can be at least partially overcome by cold-moist stratification. The seed coat of some species also may retard germination.

Acid scarification of seeds of Pacific Serviceberry (*Amelanchier laevis*) in concentrated H_2SO_4 is recommended. Cold-moist stratification requirements are 2–6 months at 41°F.

Serviceberry seeds may be sown in the fall or stratified seeds sown in the spring.

Ambrosia dumosa—Burrowbrush

Burrowbrush is abundant on well-drained soils in the deserts of the American Southwest. By virtue of its rhizomatous growth habit, *Ambrosia* is an extremely long-lived shrub. It is a potential species for environmental plantings in desert areas.

Burrowbrush is a member of the sunflower family. The fruit is a bur with spines. The burs are used as seed. Some germination will occur without pretreatment, but 30 days' cool-moist stratification at 34°F markedly improves germination.

Amorpha—Amorpha, False Indigo

In North America, the false indigos include about 15 closely related species of deciduous shrubs or subshrubs. Some species die back almost to the ground nearly every year. The false indigos provide wildlife food and cover; and some species, because of their handsome foliage and flowers, are suitable for environmental plantings.

Amorpha belongs to the legume family, so the fruit is a pod. The pod is indehiscent, gland-dotted, and contains 2 glossy seeds.

The ripe pods can be stripped from the branches and should be spread out in thin layers to permit drying.

Extraction of seeds is not necessary as the pods do not inhibit germination. No pretreatment is necessary for fresh seeds of California Amorpha (*A. californica*) and also for fall-sown seeds of Leadplant Amorpha (*A. canescens*). Stored seeds of these two species and the seeds of all other species have impermeable seed coats. Hot-water soaking for 10 minutes has improved germination.

Pods are usually sown on beds in the fall and covered with ¼ inch of soil.

Aralia—Aralia

The genus *Aralia* is comprised of about 20 species of trees and shrubs found in North America, Asia, and Australia. They are used for ornamental purposes and wildlife feed.

The fruit is a small, berry-like drupe containing 2–5 light-reddish-brown nutlets. Fruits can be stripped from the plants, and seeds recovered by maceration and flotation. Seeds of most species require scarification and cool-moist stratification.

Arctostaphylos—Manzanita

Arctostaphylos includes about 50 species of woody evergreen plants varying from low prostrate shrubs to small trees. Manzanitas occur from Central America through North America. Characteristically, the manzanitas have crooked branches with thin, usually smooth red or brown bark that exfoliates. The manzanita species offer great potential for use in native plant gardens. Use of these species has been limited because the seeds are so difficult to germinate.

Small white or pink flowers bloom in the early spring. The fruit is berry-like with granular pulp. The fruits ripen from summer to fall and have 4–10 stony seeds which may be separate or united in various numbers depending on the species.

Fruit may be collected by hand from the plants or picked up off the ground. The color of ripe fruits varies from bright red or pink to dark brown depending on the species.

The outer fleshy part of the fruit may be removed by macerating the fruits with water and separating the nutlets by flotation or air-screening. Seeds of *Arctostaphylos* species usually have hard seed coats and dormant embryos. The recommended germination enhancement treatments for Bear Berry (*A. uva-ursi*) include acid scarification in H_2SO_4 for 3–6 hours followed by warm-moist stratification at 80°F for 60–120 days, and finally cold-moist stratification at 41°F for 60–90 days. All of these treatments collectively produced 50% germination.

Good germination has been obtained for seeds of some species of *Arctostaphylos* by soaking them for 2–5 hours in H_2SO_4 and sowing them in the ground in early summer. The seeds germinate the following spring. Seed beds should be mulched over winter.

Aronia—Chokeberry

The chokeberries are deciduous shrubs native to eastern Canada and the northeastern United States. All species of this member of the rose family are valuable sources of food for wildlife during the fall and winter. The handsome foliage, flowers, and fruits of the *Aronia* species make them attractive as ornamentals, but none have been cultivated extensively.

The fruits of chokeberry are dry, berry-like pomes, containing 1–5 seeds. Natural seed dispersal is chiefly by animals.

To avoid the loss of fruits to birds, the chokeberry fruits should be picked as soon as they ripen. The red-fruited species can be left on the bushes longer than black-fruited species.

Commercial seed usually consists of the dried pomes, which are listed in seed catalogs as "dried berries." Seeds may be extracted by standard maceration techniques and flotation, followed by drying and air-screening.

Chokeberry seeds have an internal dormancy that has been broken by cool-moist stratification at 32°–41°F for 90–120 days.

A common nursery practice is to soak dried fruits in water for a few days and mash the pulp. The entire mass is stratified until spring. Some growers limit the stratification period to 60 days for Purple (*Aronia prunifolia*), 90 days for Red (*A. arbutifolia*), and 120 days for Black Chokeberry (*A. melanocarpa*). As a rule of thumb, one pound of cleaned seed may yield about 10,000 usable plants.

Artemisia—Sagebrush, Wormwood

The genus *Artemisia* has species distributed throughout much of the Northern Hemisphere and into South America. Over 100 species of herbs and shrubs that are usually aromatic compose the group. In the western United States the woody species of sagebrush were the natural vegetation on at least 96 million acres. In the Intermountain and Pacific Northwest regions, there are over 20 recognized species, subspecies, and forms of woody sagebrush. Until recently, seeds of the sagebrush species rarely entered the seed trade. Within the last decade, various types of sagebrush seeds have been specified for use in wildland revegetation projects. Collections from native stands often enter the seed trade as highly variable mixtures of flower parts, fragmented inflorescences, leaves, and, hopefully, seeds.

Recently, scientists have refined sagebrush taxonomy. Older classification systems lumped the sagebrush material together with a relatively few distinct species. This material has been shown to consist of several species and many subtaxa, each with its own morphology, genetic makeup, and ecological requirements. Many of the taxonomic separations of sagebrush have marked economic importance for the seed industry, which underscores the necessity for accurate seed identification.

The most important sagebrush seeds in commerce are probably those of Basin (*A. tridentata* subsp. *tridentata*), Mountain (*A. tridentata* subsp. *vaseyana*), and Wyoming (*A. tridentata* subsp. *wyomingensis*) Big Sagebrush. The three subspecies of Big Sagebrush grow in and are adapted to relatively distinct environments, hence the importance of identifying seed sources for revegetation.

Most *Artemisia* species flower in late summer. There are exceptions, such as Budsage (*A. spinecens*), which flowers in the early spring. The seeds, which are very small achenes, mature in the fall or early winter. Usually the entire inflorescence is stripped by hand. This results in very trashy seed collections. The aromatic flower parts of sagebrush species can induce allergenic reactions from susceptible individuals.

Seeds of some species of sagebrush have been reported to have afterripening requirements. Apparently after these requirements are satisfied, no pregermination treatments are necessary to obtain satisfactory germination.

Asimina—Pawpaw

Species of *Asimina* provide food for wildlife. Small Flower Pawpaw (*A. parviflora*) is native to the southeast, and Pawpaw (*A. triloba*) is widely distributed in the eastern United States.

Pawpaw flowers are solitary, perfect, and greenish purple. The fruits are fleshy berries that contain several dark brown, shiny seeds. Fruits are greenish yellow before maturity and turn brown to black as they ripen in August and September. The seeds are oblong, rounded, flat, and bony. *Asimina* fruits should be picked or shaken from the trees as soon as the flesh is soft. The seeds may be extracted by macerating the fruits in water and floating off the pulp. The entire fruit can be sown. Germination is usually very slow because seeds have dormant embryos and seed coats are slowly permeable. Cool-moist stratification for 60 days at 41°F has been recommended. Stratification for 100 days may be necessary.

Atriplex—Saltbushes

Atriplex constitutes an essentially cosmopolitan genus of more than 100 species, many of which are economically important to agriculture. *Atriplex* species are very important as wildlife cover and food plants. Many of the species furnish nutritious and preferred browse for domestic livestock, and most of the species are adapted for growth in saline-alkaline soils or extremely arid situations.

The female flower of *Atriplex* species has no perianth, protection being provided by two bracteoles; the bracteoles enlarge and enclose the seed, forming a false fruit or utricle. The utricle is usually considered to be a seed in the commercial seed industry.

Fully mature fruits can be shaken or hand-stripped from branches. In commercial operations, vacuum seed harvesters have been used. Before sowing, the seeds of many *Atriplex* species are run through a hammer mill to remove the enlarged portions of the utricle. This serves two purposes. First, the removal of the bracts reduces the volume of seed that must be handled and transported. Secondly, the bracts contain accumulations of soluble salts, which often require leaching before germination can occur. Care must be taken in all hammer mill treatments not to injure the seed.

The seeds of many saltbush species have afterripening requirements that must be satisfied before germination can occur. The time required for those requirements to be satisfied varies from one month to a year, depending on the species. Germination of many species of

Atriplex can be increased by presoaking the fruits and then wringing the water and dissolved salts from the material.

In a recent study of the germination of seeds of Gardner Saltbush (*A. gardneri*), Ansley and Abernethy subjected seeds collected in the Red Desert of Wyoming to pregermination treatments of washing, scarification, and/or stratification to alleviate dormancy. In the laboratory, all treatments were equal in enhancing germination. In the field, seeds that were scarified and stratified before planting produced the best seedling establishment.

When purchasing seeds of wildland species of *Atriplex,* care should be taken to identify the source of the seeds. Seeds collected at low elevations in the southwest will not be winter-hardy when planted at northern locations.

Baccharis—Baccharis

The genus *Baccharis* is composed of about 250 species of deciduous or evergreen shrubs and herbs native to North and South America. *Baccharis* plants are of little value for browse, and some species are poisonous to livestock. Some species are used as ornamentals, some for erosion control, and some for medicinal purposes.

The female flowers of *Baccharis* develop into compressed achenes which are tipped with a pappus of bristly hairs.

The ripe fruits of *Baccharis* can be collected by hand or by brushing them onto cloth or plastic sheets spread beneath the shrubs. The fruits can be rubbed to remove the pappus.

No pregermination treatments are necessary for *Baccharis* seeds. Seeds may be sown in the fall or early spring in flats or seed beds using a sandy soil mixture.

Beloperone californica—Chuparosa

A member of the acanthus family, Chuparosa is found along watercourses in the creosote bush scrub deserts of the American Southwest and adjacent Mexico.

The fruit is a capsule. Seeds germinate without pretreatment.

Berberis—Barberry, Mahonia

The barberries include about 280 species of evergreen or deciduous, spiny, or unarmed shrubs native to Asia, Europe, North Africa, and the Americas. Some authorities place about 90 of these species (all evergreen, unarmed plants) in a separate genus, *Mahonia.* Because of their handsome foliage and often attractive flowers or fruits, many of the barberries are grown for ornamental purposes. The barberries are also of value for wildlife food and cover. The majority of barberries are an alternate host for the black stem rust of wheat.

The fruit of *Berberis* species is a berry with one to several seeds. Seeds are dispersed by birds and small mammals. Ripe barberry fruits may be picked by hand, using heavy gloves, or they can be flailed onto cloth or receptacles spread beneath the bushes. The ripe fruits may be run through a macerator with water, and the pulp floated off.

There is considerable variability among species of *Berberis* in the degree of seed dormancy. The seeds of some species germinate without pretreatment while others are so dormant they fail to germinate after enhancement treatments. Cold-moist stratification is recommended for seeds of dormant species. Under natural conditions, barberry seeds germinate in the spring following seed dispersal.

Clean seed is preferred for sowing in nurseries because of mold problems with berries.

Brickellia—Brickle Bush

Brickellia is a genus of the sunflower family that contains about 100 species in North and South America. California Brickle Bush (*B. californica*) is found in many plant communities from California to Texas and Mexico. This species has potential for native plant gardens in the Southwest.

The fruit is an achene that germinates without pretreatment.

Bumelia lanuginosa—Gum Bumelia

Gum Bumelia is a spiny shrub or small tree widely distributed across the southern United States. It has some value as a wildlife species and has been planted as an ornamental.

The fruit of Gum Bumelia is a single-seeded drupe. It turns purplish black as it ripens in September and persists on the shrub into winter.

Fruits should be picked as soon as they are mature. The flesh can be removed by maceration in water.

Gum Bumelia seeds require scarification in H_2SO_4 for 20 minutes. Following scarification, 4–5 months of stratification at 35°–45°F is necessary to obtain marked germination.

Calhaudra eriophylla—Fairyduster

Fairyduster is a member of the legume family. This densely branched shrub is found from far southern California and adjacent Mexico to Texas in the creosote high zone.

The fruit is a long, linear legume. The seeds are roundish to ovate. The seed coats are hard and require a hot-water treatment for germination.

Calycanthus occidentalis—Sweet Shrub

This aromatic deciduous shrub is one of four species native to North America. They are often cultivated for their pleasantly aromatic leaves.

The fruit consist of many one-seeded achenes completely enclosed by the receptacle. For fresh seeds, no pretreatment is required for germination. More uniform germination is obtained by cold-moist stratification.

Campsis radicans—Common Trumpet Creeper

Common Trumpet Creeper, a deciduous vine, is native from Texas and Florida north to Missouri, Pennsylvania, and New Jersey. The vine is an important source of wildlife food.

The fruit of Common Trumpet Creeper is a two-celled, flattened capsule about 2–6 inches long. The capsule contains small, flat, winged seeds. Ripe capsules should be gathered before they split open after they turn grayish in the fall.

Cold-moist stratification for 60 days at 41°–50°F is recommended.

Caragana arborescens—Siberian Pea Shrub

Siberian Pea Shrub or Caragana was introduced to the United States from Manchuria and Siberia. This member of the legume family is widely used for conservation plantings on the northern Great Plains.

The fruit is a pod that contains about 6 reddish-brown seeds. The optimum seed-collecting period for Siberian Pea Shrub fruits is less than 2 weeks. The pods must be gathered by hand as soon as the pods are ripe.

Germination has been improved by cool-moist stratification for 15 days at 41°F.

Carpenteria californica—Carpenteria

Carpenteria, also known as Bush Anemone, is an erect evergreen shrub with a restricted distribution in the southern Sierra Nevada Mountains. Carpenteria has been cultivated since 1880 and is a valuable ornamental species.

The fruit of Carpenteria is a leathery, beaked capsule containing many small seeds.

Germination without treatment is excellent.

Cassiope mertensiana—White Heather

The white heathers are small creeping or prostrate alpine shrubs. They belong to the heath family. The genus includes several species of circumboreal distribution.

The fruit is a capsule. The seeds are minute and winged. Germination is enhanced by cold-moist stratification.

Castanopsis—Chinkapin

The genus Castanopsis includes about 30 species of evergreen shrubs. Golden Chinkapin (C. chrysophylla) and Sierra chinkapin C. sempervirius) often reach tree size. The chinkapin fruit consist of 1–3 nuts enclosed in a spiny bur. The nuts mature in the fall of the second year. Fruits should be hand-picked in late summer or early fall before the burs open. Germination of chinkapin seeds is not particularly good. Cool-moist stratification has not improved germination.

The nut that is contained within the bur is edible. These shrubs offer potential for use in native plantings in such difficult habitats as granitic soils at Lake Tahoe in the Sierra Nevada.

Cassia armata—Armed Senna

Armed Senna is a member of the legume family. It is a common shrub in the Mojave and Colorado Deserts of the American Southwest.

Dry pods can be hand-picked and threshed with the seed recovered by air-screening. As is the case with the seeds of many legume species, the seeds of Armed Senna have hard seed coats. Acid scarification greatly enhances germination.

Ceanothus—Ceanothus

The genus *Ceanothus* consists of about 55 species native to North America. Most species occur along the far Pacific slope with only two found naturally east of the Mississippi River. *Ceanothus* species may be evergreen or deciduous shrubs. Many species are used as ornamentals. Species of *Ceanothus* are very important as browse species for big game animals. The roots of most species of *Ceanothus* bear nodules which contain organisms that fix atmospheric nitrogen. The nitrogen enrichment that results from growth of nodulated *Ceanothus* plants may be very important in the ecology of range and forest sites.

The fruit of *Ceanothus* plants is a 3-lobed capsule bearing smooth-coated seeds. As the capsules split, the seeds are ejected with considerable force. Seed collection requires tieing cloth or paper bags around clusters of green seed capsules. As an alternative, tarps or sheets can be spread on the ground. This is always difficult to do in wildland situations. The capsules of some species can be clipped when green and allowed to open on a tarp or in a net bag.

The seeds of most species of *Ceanothus* are quite dormant. Many of the species of *Ceanothus* are fire species in which the seeds lie dormant in the soil for years, even centuries, before the hard seed coat is ruptured by the heat of a passing wildfire. Hot water treatments—dropping seeds into 200°F and leaving them in water until they cool, have been substituted for this hard seed coat germination requirement. Embryo dormancy may still be a problem after the hot water treatment. Cold-moist stratification enhances germination of seeds of many species of *Ceanothus*.

Celastrus scandens—American Bittersweet

American Bittersweet is a deciduous climbing or twining shrub of eastern North America. The plant is valuable for ornamental purposes, and some food and cover.

The light reddish seeds of American Bittersweet are borne in capsules. The seeds are surrounded by fleshy arils. The seeds should be collected as soon as the capsules split, exposing the aril.

Seeds of American Bittersweet have a dormant embryo that requires a period of afterripening. Good germination is obtained from fall seeding or by spring sowing of seeds stratified at 41°F for 2–6 months.

Cephalanthus occidentalis—Common Buttonbush

Common Buttonbush is a deciduous shrub widely distributed in North America from New Brunswick to California. The seeds of this species are eaten by many birds, and the species is a bee plant for honey production.

The fruit of Common Buttonbush is composed of 2 or 4 single-seeded nutlets joined at their bases.

Fruit collection can begin as soon as the buttons turn reddish brown. Common Buttonbush seeds germinate promptly without pretreatment.

Ceratoides—Winterfat

Ceratoides (formerly *Eurotia*) species inhabit rangelands in western North America and Central Asia. Winterfat (*C. lanata*) is one of the most important browse species on the western range. Growing at relatively low elevation sites located between sagebrush and salt desert ranges, winterfat has been used as a winter forage for cattle and sheep. Much of the natural stands of winterfat have been destroyed by excessive grazing. Eurasian Winterfat (*C. latens*) serves a similar grazing purpose in Central Asia.

The fruits of winterfat ripen in October. The fruits, which are one-seeded utricles, are borne in compound spikes. The utricles consist of a nutlet enclosed in two bracts which bear fluffy white hairs. The curved embryo almost completely encircles a very thin layer of residual endosperm. The fruits of winterfat can be mechanically collected with a seed stripper. Manual stripping is also feasible.

Freshly collected seeds germinate without pretreatment. The seeds do not have a long storage life. As is the case with seeds of several species in the goosefoot family, the germination of stored seeds of winterfat may be very low.

The germination of seeds of winterfat and Eurasian Winterfat was compared at 55 constant and alternating temperatures. Seeds of both species germinated at a wide range of temperatures. Optimum germination occurred most frequently at 32°–40°F nighttime temperatures alternating with 50°–70°F daytime temperatures. There were large differences among commercial sources of winterfat seeds in germination.

The acute end of winterfat seeds is formed by the apex of the radicle and of the cotyledons. Apparently this shape makes embryo root caps especially susceptible to damage during threshing, and such damage is the cause of a high percentage of seedlings lacking geotropism or proper directions of root growth. Entire fruits rather than threshed seeds should be seeded by broadcasting.

Although winterfat offers much promise for range improvement, it has not been extensively used because of erratic seedling establishment.

Cercidium floridum—Palo Verde

Palo Verde is a large shrub or small tree with smooth green bark. This member of the legume family is leafless most of the year. *Cercidium* is a North American genus of about 10 species. The fruit is a several-seeded legume. The smooth green bark of this desert shrub makes it an attractive prospect for desert gardens.

No treatment is necessary for seed germination.

Cercis—Redbud

Species of Redbud occur in North America, Europe, and Asia. With brilliant pink to reddish-purple flowers and distinctive flattened pods for fruits, the species of *Cercis* are unique shrubs in flower or fruit.

As a member of the legume family, the fruit of the *Cercis* species is a large two-valved pod. Seeds are small, brown, and hard. Pods should be collected as soon as they dry to reduce insect damage.

Seeds of Redbud have hard, impermeable seed coats in addition to internal dormancy. Both scarification and cold stratification are needed. Acid scarification or hot water treatments have been used, followed by 1–3 months' cold-moist stratification.

Cercocarpus—Mountain Mahogany

The Mountain Mahogany group consist of about 10 species of evergreen shrubs native to western North America. As members of the rose family, the *Cercocarpus* are considered important browse species on rangelands.

The fruit of *Cercocarpus* species is a cylindrical achene with a persistent feathery style 2–3 inches long. The achene and style are covered by short white hairs that can irritate the skin.

Seeds should be collected as soon as the fruits are dry. This is usually indicated by a white aspect to the fruiting branches caused by the style hairs drying. The seeds can be stripped or flailed into a container or onto tarps. Avoid exposure of skin to the hairy styles.

True Mountain Mahogany (*C. montanus*) seeds may germinate without pretreatment, while seeds in some lots may have varying degrees of dormancy. Seeds of Birchleaf Mountain Mahogany (*C. betuloides*) germinate without pretreatment. Seeds of Curlleaf Mountain Mahogany (*C. ledifolius*) are almost always very dormant. Germination of seeds of this species can be obtained by cold-wet stratification in an aeriated water bath. Seed of some lots of Curlleaf Mountain Mahogany can be induced to germinate by soaking for 2–5 hours in a 5% solution of H_2O_2.

Cereus giganteus—Saguaro

Saguaro is the largest of the columnar cacti growing naturally in the United States. Its natural range is restricted to the Sonoran Desert in the Southwest and Mexico. This is a prime species for desert gardens in the Southwest.

The green fruits turn red when ripe. Each fruit contains about 2,500 black seeds. The fruit pulp must be thoroughly removed by washing.

Seeds require near red, or shorter wave-length, light for germination. A 77°F incubation temperature with 8 hours of light daily usually gives excellent germination.

Chamaebatia foliolosa—Bearmat, Mountain Misery

Bearmat is a dense-growing shrub found on the west slopes of the Sierra Nevada and the mountains of Baja California. A number of the rose family, the fruit of bearmat is a brown achene. This species has potential for ground cover in forested situations.

Seeds require 1–3 months' cold-moist stratification at 35°–41°F before they will germinate.

Mountain misery is a shrub belonging to the rose family that forms dense low-growing patches in conifer woodlands on the west slope of the Sierra Nevada in California.

The fruit is an achene. The achene has pronounced embryo dormancy which requires 3 months of cold-moist stratification.

Chamaebatiaria millefolium—Fernbush

Fernbush is a monotypic representative of the rose family found growing on the east slope of the Sierra Nevada and across the northern intermountain area to Wyoming and south to the mountains of northern Arizona. This unusual species is worthwhile to grow as a specimen.

The fruit is a leathery follicle. No pregermination treatment is required for fresh fruits. Stored seeds require 3 months' cold-moist stratification.

Chilopsis linearis—Sweet Desertwillow

Desertwillow, which is a member of the bignonia family, grows along dry washes in the arid Southwest. The plant is useful for wildlife cover, erosion control, and ornamental plantings in arid regions.

The fruit is a two-celled capsule. It ripens from late summer to late fall and persists through the winter. Seed pods can be hand-picked after late September and through the winter. As the fruits dry, the light brown seeds that are fringed with soft hairs can be shaken loose.

Desertwillow seeds are not dormant, but storage in wet sand will speed germination.

Chrysothamnus—Rabbitbrush

Rabbitbrush species are members of the sunflower family. This group of highly variable plants constitutes one of the most important types of shrubs on sagebrush rangelands. Most of the Rabbitbrush species sprout from their crowns after the aerial portions of the shrubs are burned in wildfires. Because most sagebrush species do not sprout when burned, the subdominate Rabbitbrush species characterize sagebrush communities for several years after wildfires. Rabbitbrush species are not restricted to sagebrush rangelands. Species of Rabbitbrush occur in the understory of conifer woodlands and in the moister portions of the salt desert. In certain situations, ecotypes of Rabbitbrush species are useful browse species, but the browse of most species is avoided by large herbivores. Volatile chemicals from the foliage and the wind-blown pollen of Rabbitbrush species can be very allergenic.

The value of Rabbitbrush species is their ability to establish and grow in environments where few other perennial native plants have a chance. There are two major Rabbitbrush species: Gray Rabbitbrush (*C. nauseosus*) and Green Rabbitbrush (*C. viscidiflorus*). Each species is composed of a host of subspecies, varieties, forms, and ecotypes, the extent of which is limited only by the patience of the collector.

Rabbitbrush species flower in late summer. In dense stands, the masses of yellow flowers characterize the landscape. The seed is an achene crowned with a ring of pappus hairs. The seeds are dispersed by wind as soon as they mature. Seeds can be stripped from the shrubs or shaken onto tarps.

Rabbitbrush seeds are not dormant for most collecting, but the viability is often quite low. For many collections the achenes are largely aborted fruits.

Direct seeding of seeds of Rabbitbrush species is not often successful. Seedlings can be easily transplanted. An easy way to propagate seedlings is to crumble inflorescences containing ripe achenes on the top of flats of greenhouse soil and cover the seeds with a thin layer of vermiculite.

Clematis—Clematis

The genus *Clematis* includes more than 200 species of climbing vines or semi-shrubs widely distributed in the Northern Hemisphere. Many horticultural varieties are grown for ornamental purposes.

Clematis fruits are borne in heads of one-seeded achenes with persistent feathery achenes. Fruits are brown when ripe and may be gathered from the plants by hand.

Clematis seeds have dormant embryos and require cold-moist stratification at 33°–40°F for 60–180 days to obtain germination.

Coleogyne—Blackbrush

Blackbrush characterizes a vegetation zone across the southern intermountain area between pinyon/juniper woodlands and true desert plant communities. A member of the rose family, blackbrush has achenes for fruits. The villous styles persist on the achenes. Seeds require cool-moist stratification for germination.

Colubrina californica

C. californica is an intricately branched shrub of the buckhorn family. It is found occasionally in dry canyons in the Creosote Bush zone of Southwestern deserts. There are about 18 species of *Colubrina* in North and South America.

The fruit is a capsule. There is very little information available concerning the germination of this species, but apparently the seeds will germinate without pretreatment.

Colutea arborescens—Common Bladder-Senna

Common Bladder-Senna is a member of the legume family introduced to North America for conservation plantings. Common Bladder-senna fruits are indehiscent inflated pods. The pods ripen from July to October and contain several small seeds. Ripened pods can be hand-picked when ripe and threshed to recover the seeds.

The seeds have hard seed coats and must be acid-scarified to obtain marked germination. Concentrated H_2SO_4 is recommended with the duration of treatment, dependent on the seed lot.

Comarostaphylis diversifolia—Summer Holly.

Summer Holly is an erect evergreen shrub found in the coastal chaparral of southern California and adjacent Mexico. This member of the heath family has red drupes for fruits. The seeds require 3 months' cold-moist stratification for germination.

Cotoneaster—Cotoneaster

The cotoneaster include about 50 species of deciduous or evergreen shrubs native to Europe, Africa, and Asia. The more hardy species are used commonly in mass plantings, hedges, and shelter belts.

The fruits are small, black or red berrylike pomes which ripen in late summer and persist into the winter. Each fruit contains 1–5 seeds. The fruits should be collected by hand and the seeds extracted by macerating with water.

The seeds of cotoneaster species have hard seed coats and embryo dormancy. The seeds require acid scarification followed by cold-moist stratification for marked germination.

Cowania mexicana var. *stansburiana*—Cliffrose

Cliffrose is another member of the rose family that is a valuable browse species on the western range. Native to the southwestern United States and adjacent Mexico, Cliffrose stands extend into the southern portions of the Great Basin.

The white to sulfur-yellow flowers of Cliffrose make the shrubs one of the most attractive species of the major browse plants. The fruit is an achene containing a single seed. The achene has a long feathery style which usually falls off. The first seeds to set are usually of the highest quality. They can be collected by hand or with vacuum equipment.

Without pretreatment, germination of Cliffrose seeds is usually quite low. Cold-moist stratification is required to enhance germination.

In nature, most Cliffrose seeds are collected by rodents and cached in groups in the soil. The seeds receive natural stratification during the winter, and germination occurs in the spring. In artificial seeding, rodents will often collect and cache sown seeds of Cliffrose.

Crossosoma californicum—Crossosoma

This genus belongs to the obscure crossosoma family. Crossosoma is found on the California channel islands in coastal sage communities. The fruit is a follicle. The seeds have a conspicuous fringed aril. No pre-germination treatment is required. A relatively rare specimen-type species.

Cytisus scoparius—Scotchbroom

Scotchbroom is an evergreen shrub that is a member of the legume family. This native of Europe has escaped from cultivation and is considered a serious pest along the east and west coast of North America.

The fruit is a narrow, oblong pod containing brownish-black seeds. Seeds of *Cytisus* have hard seed coats that require scarification for marked germination. Hot water or H_2SO_4 scarification has been recommended. The seed coats can be mechanically clipped to promote germination.

This species can be a pest and is subject to a biological control program.

Dalea (now Psorothamnus)—Indigobush

Dalea consists of over 100 species of legumes found in North and South America. The stems are usually gland-dotted. Smoketree (*D. spinosa*) is found in the creosote bush communities of the Mojave Desert in California and Arizona. Some of the species in the indigobush group have great potential for native plant gardens. The legume-type flowers are very striking and the strange, gland-covered stems very unusual.

The fruit of the indigobush is a short, ovoid legume with 1 or 2 seeds. Seeds require no pretreatment for germination.

Dendromecon rigida—Stiffbush Poppy

Stiffbush poppy is an openly branched, evergreen shrub native to California and Baja California. Fruits are linear capsules that separate incompletely at maturity. Ripe fruits may be collected in May through July. Seeds are black.

Little is known about the germination of this species, but plants have been obtained by incubating the seeds at an alternating temperature regime of 40°–70°F for 50 days. Germination can be enhanced by burning litter on the surface of flats after the seeds are seeded.

Encelia virginensis—Virgin River Encelia

On desert slopes, dry washes, and ravine embankments, Virgin River Encelia is found in the Colorado and Mojave Deserts of the American Southwest. This member of the sunflower family has potential for environmental plantings. Hand-harvesting is the only practical method of collecting the achenes of this species. Achenes germinate without pretreatment.

Ephedra—Ephedra

The genus *Ephedra* is widely distributed in the Northern Hemisphere. The *Ephedra* belong to the order Ephedrales of the Coniferophta, or cone-bearing plants; as such, they are relatives of the conifers. Green (*E. viridis*) and Gray (*E. nevadensis*) are widely distributed in the Intermountain Area of the western United States. Their unique foliage and attractive evergreen or gray color make them desirable species for environmental plantings.

On good seed years abundant collections of ephedra seeds can be obtained by flailing the fruiting branches over an open tray.

The seeds germinate best at alternating temperature requirements with quite cold nighttime temperatures. Seedlings grow rapidly and can be easily transplanted.

Epigaea repens—Trailing Arbutus

Trailing Arbutus is an evergreen, prostrate, creeping shrub that grows natively in the eastern United States and Canada. The fruit is a 5-lobed, hairy, dehiscent capsule. The seeds are imbedded in a fleshy pulp. Capsules may be collected after they mature but before they eject the seeds.

The seeds are very small, and after separating from the pulp are best propagated by sprinkling on the surface of greenhouse soil mixtures in pots or trays.

Eriogonum—Buckwheat

Eriogonum is a genus of more than 200 species in the polygonum family. Many of the species are herbs, but important species of buckwheat are shrubs or half-shrubs. Buckwheat species with low-growth forms, attractive foliage, and showy flowers are promising species for environmental plantings.

The fruit of buckwheat species is a three-sided achene. The achenes can be hand-stripped from the plants. The achenes generally germinate without pretreatment. In such a large genus, one would expect a variety of germination responses.

Euonymus occidentalis—Western Burning Bush

Deciduous shrub with slender branches and smooth greenish bark. Found on damp wooded banks along streams in the coastal forest from California to Washington.

The seeds are surrounded by a red aril. Seeds are dormant and require three months' stratification.

Fallugia paradoxa—Apache Plume

Apache Plume is an attractive shrub native to the southwestern United States and adjacent Mexico. This member of the rose family is closely related to both Cliffrose and Bitterbrush. Apache Plume is a valuable browse species, and its attractive foliage and plumed seeds make it desirable for environmental plantings.

The fruit is a hairy achene with a persistent feathery style 1–2 inches in length. The fruits usually occur in dense clusters of 20–30.

Seeds may be collected when the reddish color of the hairy styles whitens and the plump seeds fall readily. Seeds are reported to germinate without pretreatment at low incubation temperatures.

Fremontodendron—Fremontia

The fremontia are handsome arborescent shrubs with brilliant flowers that make them desirable for environmental plantings. They are native to the southwestern United States.

The fruit is a dense, wooly or quite bristly, egg-shaped capsule containing numerous brown seeds. For the species with bristly fruits, care must be taken in harvesting. Collect when the capsules begin to split.

Fremontia seeds require scarification in hot water followed by cold-moist stratification at 35°F for 12–16 weeks to obtain marked germination.

Fouquieria splendens—Ocotillo

The Ocotillo or Candlewood plants are resinous shrubs found in the Southwest and adjacent Mexico. The fruit is a capsule. The seeds are oblong, compressed with wings. The wings break into hairlike parts.

This species has potential for use in desert gardens.

Little is known about the germination of seeds of these shrubs. Fritz Went observed natural seedling germination under parent shrubs following a summer rain.

Galvezia speciosa—Bush Snapdragon

Bush Snapdragon is a bright green shrub found in rocky canyons on the channel islands of California. About four species of Galvezia occur from California to Peru. These species have great potential for use in native plant gardens. The fruits are capsules. The seeds are cylindric with irregular winglike plates. Seeds germinate without pretreatment.

Garrya—Silktassel

The genus *Garrya* consists of four species native to far western North America. The silktassel fruit is a rather dry one or two-seeded berry that ripens to a dark purple in the fall. Ripe fruit can be gathered by stripping. Seeds can be recovered by macerating the fruits in water.

Germination of silktassel seeds is variable depending on the species and source, but most appear to require cold-moist stratification. Stratification at 36°–41°F for 30–120 days has enhanced germination.

Gaultheria—Wintergreen

The genus *Gaultheria* contains about 100 species with nearly worldwide distribution. The species provide food and cover for a large number of wildlife species. Oil of wintergreen is extracted from the leaves of some species.

The fruit of *Gaultheria* species is a many-seeded capsule surrounded by a pulpy calyx that forms a pseudoberry. Depending on the species, the fruits can be combined, stripped, or picked individually from plants. Either dry milling or maceration can be used to extract seeds.

Most species of wintergreen require cold-moist stratification, but the duration of the requirement varies among species.

Gaylussacia baccata—Black Huckleberry

Black Huckleberry is a small deciduous shrub from Louisiana east to Florida and northward to Maine and Manitoba. The berries are an important game feed. The black berrylike, drupaceous fruit mature from July to September. Fruits may be stripped from branches by hand or with a blueberry rake. Seeds may be extracted by macerating the berries in water.

Untreated seeds are slow to germinate. Warm stratification for 30 days followed by incubation with cold temperature has enhanced germination.

Grayia—Hopsage

Spiny hopsage (*G. spinosa*) is a widely distributed shrub on sagebrush rangelands in the western United States. It is considered a valuable browse species. Members of the goosefoot family, the fruits of species of *Grayia* are utricles consisting of a nutlet enclosed by two flattened papery bracts. The fruits mature in June and July, and when tinged with red are very attractive. Fruits can easily be stripped from the plants. It is not necessary to extract the seeds.

The seeds germinate without pretreatment. This is one of the few native shrubs that consistently produce highly germinable seeds.

Seedlings of Spiny Hopsage require chilling for bud burst and growth after initial seedling growth.

Haplopappus—Goldenweed

The genus *Haplopappus* of the sunflower family consist of about 150 species of herbs, half-shrubs, and shrubs native to the Western Hemisphere. Similar in appearance to Rabbitbrush species, the woody species of *Haplopappus* are widely distributed in a number of rangeland plant communities. Goldenweeds are usually considered to be weeds.

The fruit of *Haplopappus* species is an achene which often is covered with silky hairs. Pappus hairs or bristles may persist or dehisce.

Seeds usually germinate without pretreatment.

Hippophae rhamnoides—Common Sea Buckthorn

Native to Europe and Asia, Common Sea Buckthorn is planted in the United States as hedges and screens. The species provides cover and food for wildlife.

The fruit is acidic, orange-yellow, and drupelike, about the size of a pea. The fruit ripens in the fall. Fruits persist during the winter and may be picked until spring. Prompt maceration is best to prevent acquired seed dormancy.

For marked germination, the seeds require cold-moist stratification for 90 days at 36°–41°F. Untreated seeds may be sown in the fall.

Holodiscus discolor—Oceanspray

Native to the Pacific Northwest and adjacent Canada, Oceanspray is widely found as a successional species in openings in a variety of forest types. Its range extends into the mountains of the Great Basin. In the Pacific Northwest this is an obvious species for use in native plantings. This species may be adapted for use in rock gardens in the Intermountain Area.

The fruits are light yellow achenes. The seeds require extended periods of cold-moist stratification for germination.

Hymenoclea salsola—White Burrowbush

White Burrowbush is commonly found on sandy sites of desert

washes, alluvial fans, and alkaline flats in the deserts of the American Southwest. This species should be tried in native plant gardens.

The pollen from White Burrowbush is said to cause allergies. White Burrowbush is a member of the sunflower family and has achenes for fruits.

Achenes can be collected by hand from White Burrowbush plants. The seeds germinate without pretreatment.

Hyptis emoryi—Desert Lavender

Desert Lavender is an erect aromatic shrub with numerous straight branches. *Hyptis* is a very large genus with about 350 species in the New World. Members of the mint family, the seeds are nutlets.

Little is known about the germination of these seeds. Fritz Went observed seedlings in drying washes in desert environments and assumed tumbling in the flooded washes scarified the seed coats.

Isomeris arborea—Bladderpod

Bladderpod is a drought and wind-resistant shrub native to the southern California coastal scrub. This species has been used as an ornamental species in non-irrigated plantings in the deserts of southern California.

This member of the caper family has a capsule for a fruit. Mature fruits can be hand-picked. Seeds are apparently polymorphic with white and black seeds in the same collections. The white seeds have lower germination than black seeds, but both can be germinated without pretreatment.

Kalmia latifolia—Mountain Laurel

Mountain Laurel is a broad-leafed, evergreen shrub. Its primary range is along the Appalachian Mountains in eastern North America. In the area where it is adapted Mountain Laurel makes an excellent ornamental species.

A member of the heath family, Mountain Laurel fruits are capsules. The capsules. can be picked at maturity. After the capsules dry, the seeds can be threshed and recovered by air-screening. Germination is enhanced by cold-moist stratification. Seeds should receive light during incubation.

Kochia—Kochia

Kochia consist of about 35 species of herbs or semi-shrubs mostly native to the Old World. In the salt deserts of the Intermountain Area of western North America, Green-Molly (*K. americana* is an important browse species. Prostrate Kochia (*K. prostrata*), a valuable browse species in Central Asia, has been introduced to North America as a revegetation species. Seeds of Prostrate Kochia mature quite late in the fall, when natural drying conditions are often inadequate. Artificial drying may not influence initial germination, but such drying may influence storage life of the seeds.

Members of the goosefoot family, the *Kochia* species have globose utricles for fruits. The seeds germinate very rapidly without pre-treatments. The seeds have a very short half-life in bulk storage.

Larrea tridentata—Creosote Bush

Creosote Bush is an evergreen shrub native to the arid regions of the southwestern United States, Mexico, Argentina, and Chile. Creosote Bush is the landscape-dominant species over vast areas of desert environment. This species is not normally thought of as an ornamental species, but because of its dominance in the deserts it is often destroyed in construction projects in environments where few other shrubs can be grown.

The fruit of Creosote Bush is a densely white villous capsule. When the fruits are cast they separate into individual carpels, each normally containing one seed.

Creosote bush seeds exhibit hard seed coats that require scarification to induce germination.

Lavatera assurgentiflora—Malva Rose

A member of the mallow family, this erect shrub, which is native to the California channel islands, has become naturalized in the southern coastal region of California. It has attractive rose-colored flowers. This is a valuable species for use in native gardens.

Seeds germinate without pretreatment.

Ledum glandulosum—Labrador Tea

A member of the heath family, Labrador Tea is widely distributed in the colder regions of North America and Eurasia. This is an interesting species for use in shade in mountain rock gardens. The fruit is a capsule that splits at the base. The seeds are abundant but quite minute, and elongated with wings. The seeds germinate without pretreatment. They should be germinated in the shade.

Lepidium fremontii—Bush Peppergrass

Bush Peppergrass is found in Creosote Bush and Joshua Tree woodlands communities in the American Southwest. This attractive member of the mustard family has great potential for environmental plantings. The fruit is a capsule that can be stripped from the shrubs by hand.

Seeds of Bush Peppergrass are very dormant, and germination procedures have not been developed.

Lepidospartum squamatum—Scalebroom

Scalebroom is found in the coastal sage communities and deserts of southern California and adjacent Mexico. Its unique foliage and growth form suggest its desirability for environmental plantings in desert areas.

A member of the sunflower family, Scalebroom has achenes for fruits. The fruits can be hand-picked from the shrubs. Moderate germination can be obtained without pretreatment.

Lespedeza—Lespedeza

The genus *Lespedeza* includes about 140 species mostly native to Asia. Several shrubby species have been introduced to North America. The shrubs are planted mainly for wildlife food and cover.

As members of the legume family, the fruit of the *Lespedeza* is an indehiscent pod which matures in the fall.

Shrub lespedeza seed is commonly harvested by direct combining. A high percentage of seeds have hard seed coats that require scarification. A huller-scarificer or modified hammer mill is used for this purpose. Acid scarification for 30 minutes can also be used.

Lindera benzion—Spicebush

Spicebush is native to eastern North America. It is a valuable species for wildlife food and environmental plantings. The Spicebush fruit is a red drupaceus berry ripening in late summer.

Fresh fruit should be pulped in water. Spicebush seeds have dormant embryos that respond to stratification for 30 days at 77°F, followed by 90 days at 39°–41°F.

Lonicera—Honeysuckle

The honeysuckles include about 180 species of deciduous, sometimes evergreen, upright shrubs or climbing vines widely distributed in the Northern Hemisphere. The larger species are especially valuable for wildlife and environmental plantings.

The attractive fruits of honeysuckle are paired berries that contain few to many seeds. The fruits are usually hand-picked. Extraction is accomplished by maceration.

The seeds of most species show some dormancy. Cold-moist stratification is recommended. Fall sowing usually results in good seedling establishment.

Lycium—Wolfberry

The wolfberries consist of about 100 species of the nightshade family with virtually worldwide distribution. They are used in ornamental plantings, chiefly for their showy berries.

Ripe berries may be picked from the bushes in the fall. The seeds can be extracted by maceration.

Dormancy of wolfberry seeds is variable. The seeds of some species require stratification at 41°F for 60–120 days.

Lyonothamnus floribundus—Catalina Ironwood

Catalina Ironwood is a member of the rose family native to Catalina Island. This large shrub or small tree with exfoliating bark is planted in California as an ornamental.

The fruit is a glandular pubescent follicle. Seeds germinate without pretreatment.

Menispermum canadense—Common Moonseed

Common Moonseed is a climbing woody vine native to eastern North America. The attractive fruits are valuable food for wildlife.

Fruits may be collected from September to November. Seeds may be extracted by maceration in water.

Germination is enhanced by cold-moist stratification. Stratification at 41°F for 60 days has been used.

Menodora scabra—Rough Menodora

Rough Menodora is a low shrub native to the southwestern United States. It provides useful browse for livestock and game animals. The fruit is a thin-walled capsule that ripens in the fall.

Seeds require no pregermination treatments.

Mitchella repens—Partridgeberry

Partridgeberry is an evergreen shrub native to eastern North America. This attractive plant is often used in rock gardens. Fruits are scarlet drupaceous berries that ripen in July and persist over winter. Fruits should be macerated in water to extract seeds.

Partridgeberry seeds have internal dormancy that can be overcome by stratification at 41°F for 150–180 days.

Myrica—Wax Myrtle

The genus Myrica contains about 30 species of wide distribution. Sierra Sweet-Bay (M. hartwegii) is found at mid-elevations of the west slopes of the Sierra Nevada, and Wax Myrtle (M. californica) is found in coastal communities from California to Washington. This is an interesting species to try in native plantings in situations where they are adapted.

The fruits are nutlets with a covering of resin or wax. The seeds are dormant and require three months' cold-moist stratification.

Nemopanthus mucronatus—Mountain Holly

Mountain Holly is a deciduous branch shrub that occurs in the northeastern United States and adjacent Canada. A member of the holly family, the fruit is a dull, red, berry-like drupe. Seeds can be extracted by maceration. A period of afterripening is required for embryo maturation before germination will occur.

Olneya tesota—Desert Ironwood

Olneya is a monotypic genus of the legume family. Desert Ironwood is a spinescent shrub or small tree found in creosote bush communities of the Colorado Desert of Arizona, California, and adjacent Mexico.

The fruit is a legume pod with black seeds. Fresh seeds require no pretreatment. Stored seeds should be soaked in water or may require scarification.

Osmaronia cerasiformis—Osoberry

Osoberry is a shrub native to the Pacific Northwest of the United States and Canada. Ripening fruits are highly attractive to birds such as Cedar Waxwings. The white, fragrant flowers develop into single-seed drupes for fruits. Clusters of ripe fruits can be stripped by hand. Lengthy cold-moist stratification is required to overcome seed dormancy. In one test, 120 days of stratification at 38°F were required.

Parkinsonia—Retama

The germination of seeds of the woody legume Retama (*P. aculeata*) is enhanced by scarification. Scarification in concentrated sulfuric acid for 45 minutes gave the best results.

Parthenocissus—Creeper

About 10 species and many varieties of creeper are native to either eastern Asia or North America. The most well-known species is the Virginia Creeper (*P. quinquefolia*), which is used as an ornamental species. The flowers are small, greenish, and borne in long stem clusters. The flowers produce several-seeded berries. The fruits can be hand-collected. The seeds can be extracted by macerating the fruits in water.

Natural germination takes place during the first or second spring after the seeds are produced. Germination can be enhanced by stratification at 41°F for 60 days.

Peraphyllum ramosissimum—Squaw Apple

Squaw Apple is native to sagebrush communities and juniper woodlands from central Oregon to Colorado. In locations, it is considered important deer forage.

The fruit of the Squaw Apple is a pome. The ripe fruits can be picked in June and July. The seeds can be extracted by macerating the fruits with water.

Stratification for 45 days at 38°F enhances germination.

Peucephyllum schottii—Pigmy Cedar

Pigmy Cedar is a rather openly branched aromatic shrub found in the Creosote Bush scrub of the Mojave and Colorado Deserts. *Peuce-*

phyllum is a monotypic genus of the sunflower family.

The fruit is an achene which germinates without pretreatment.

Philadelphus lewisii—Lewis Mock Orange

The natural range of Lewis Mock Orange is the Pacific Northwest. It was introduced into cultivation in the 1800s. The fruit is a capsule, which matures in late summer, with the seeds dispersed in September or October.

Seeds can be extracted by gently crushing the capsules and separating the seeds in an aspirator.

Seeds stratified for eight weeks at 41°F will give satisfactory germination.

Photinia arbutifolia—Christmasberry

Christmasberry grows in the foothills of California and into the mountains of Baja California. The attractive foliage and berries of Christmasberry make it a desirable species for environmental plantings.

Christmasberry belongs to the rose family. The fruits are pomes that turn bright red when ripe. Hand snips are needed to clip the pomes. The fruits should be placed in a warm place and quickly allowed to ferment to aid in extracting the seeds.

Christmasberry seeds will germinate without pretreatment, but germination is quite slow.

Phyllodoce—Mountain Heather

The genus *Phyllodoce* consists of about eight species with circumboreal distribution. Members of the heath family, these low-growing evergreen shrubs are found in subalpine communities. These species are important for use in rock gardens.

The fruit is a capsule that contains many minute, narrowly winged seeds. Fresh seeds germinate without pretreatment, while stored seeds may require two months' cold-moist stratification.

Physocarpus—Ninebark

Species of Ninebark form understory communities in openings in conifer forests in North America and Asia. These members of the rose family have an inflated follicle for a fruit. Ripe follicles have 2–5 shiny yellowish seeds.

Ripe fruits can be picked from the shrubs or shaken onto drop cloths.

Specific information about seed germination is scanty. Seeds can be sown in the fall for spring germination.

Porlieria—Guayacon

Seeds of Guayacon (*P. angustifolia*) germinate without pretreatment.

pretreatment.

Potentilla glandulosa—*Potentilla*

A member of the rose family, this semi-woody perennial is found in many plant communities from coastal lower California to British Columbia, and inland in the Sierra Nevada foothills and the mountains of Idaho. This could be an interesting ground-cover species in proper environmental situations.

The fruit is an achene, which can be hand-picked. The achenes are very small. Light is required during the germination process.

Purshia—Bitterbrush

Antelope Bitterbrush (*P. tridentata*) is widely distributed in the western United States, where it is among the most important browse species. Desert Bitterbrush (*P. glandulosa*) has a more restricted distribution in the southern portions of Utah, Nevada, and California.

The fruit of bitterbrush is an oblong achene. Flowers bloom in late spring, and seeds are ripe in mid-summer.

Timing of seed collection. is important to avoid seed loss. Achenes can be harvested by flailing the branches over a container. Vacuum harvesters have been used to harvest the seeds. Papery husks composed of flower parts persist around the achenes and must be removed before sowing the seeds. This can be accomplished by hand-rubbing the seeds with rubber-covered paddles or by running the seeds through specially modified hammer mills. The seeds can be separated by air-screening.

Natural bitterbrush regeneration is almost entirely rodent-connected. Rodents collect the seeds, remove the husks, and cache the seeds.

Bitterbrush seeds are quite dormant unless stratified. Stratification at 39°F for as little as two weeks will satisfy the stratification requirement.

In artificial regeneration it is very difficult to handle and sow stratified seeds because they must be sown while wet. Germination of bitterbrush seeds can be enhanced by soaking the seeds for five hours in 3 percent H_2O_2.

Rhododendron—Rhododendron

The rhododendrons comprise over 600 species of evergreen and deciduous shrubs or small trees. Many are cultivated for ornamental purposes. In North America, native species of *Rhododendron* are found in coastal and mountain forests on the East and West Coast.

The fruit of *Rhododendron* species are oblong capsules that ripen in the fall. The capsule splits when ripe, releasing large numbers of minute seeds.

Fruit collection should start as soon as the capsules lose their green color and turn brown. The seeds can be removed by threshing and air-screening to separate the chaff from the seeds.

Rhododendron seeds will germinate without pretreatment, but they require light diurnally during the germination process.

Rhus—Sumac

The sumacs include about 150 species of shrubs or small trees of temperate and subtropical regions. The handsome foliage of sumacs often turns brilliant colors in the fall.

The fruit of the sumacs is a small, smooth or hairy drupe with a single bony nutlet. Fruits may persist on the shrubs over winter. Fruit clusters may be picked by hand when ripe. The dried fruit clusters can be broken into individual fruits by rubbing or by beating in canvas sacks.

Rhus seeds germinate poorly without pretreatment. Dormancy is caused by impervious seed coats that require scarification in acid or hot water. Seeds of many species require cold-moist stratification after scarification.

Ribes—Current, Gooseberry

Ribes include about 150 species of deciduous, rarely evergreen, shrubs that occur in the colder portions of North and South America, Europe, and Asia. The smooth species are called currants, the prickly species, gooseberries. Many of the wild species have potential for use in gardens, both for ornamental value and for the fruits they produce.

The fruit of *Ribes* species is a green, many-seeded glandular or smooth berry. Fruit color varies among species. The fruits need to be picked or stripped from the bushes as soon as they are ripe to lessen losses to birds. Maceration and flotation can be used to extract the seeds.

Natural germination of *Ribes* seeds normally occurs in the spring following dispersal. Most species require at least one stratification period of fairly long duration to break embryo dormancy.

Rosa—Rose

The genus *Rosa* includes about 100 species native to Northern Hemisphere temperate zones. In the wilds, rose fruits are important food sources for numerous species of birds and mammals.

The seed of roses is an achene borne within a fleshy, berrylike fruit called a hip. The hips can be hand-picked as soon as they turn red. Achenes can be extracted by macerating the hips in water and recovering the seeds by flotation.

Stratification is required to break the dormancy of seeds of most rose species. Cold-moist stratification at 40°F for various lengths of time has broken dormancy of rose seeds.

Rubus—Blackberry, Raspberry

Rubus includes about 400 species of deciduous or evergreen, often prickly, erect or trailing shrubs. Most species are native to the cool, temperate regions of the Northern Hemisphere.

Despite the name, the fruit of the black and raspberries is an aggregate of small, usually succulent drupes, each containing a single, hard-pitted nutlet.

When ripe, Rubus fruits should be collected by hand to prevent losses to birds. The seeds can be extracted by macerating the fruits in water.

Seeds of many *Rubus* species are slow to germinate because they have hard, impermeable coats. Both H_2SO_4 and sodium hypochlorite have been used for scarification. After scarification, a complex combination of warm and cold stratification may be necessary in order to obtain germination.

Sarcobatus vermiculatus—Greasewood

Greasewood is a member of the goosefoot family that occupies and characterizes vast expanses of salt-desert vegetation in the Intermountain Area. It also occurs on saline-alkaline soils throughout much of western North America. The importance of Greasewood is that it is the only shrub that will grow in many situations where it is adapted.

Greasewood seeds consist of a spirally coiled embryo borne in a leathery utricle. The utricle has crenulate wings. Seed germination is highly variable, but best results are usually obtained by removing the seeds from the fruit.

Salazaria mexicana—Bladder Sage

Bladder Sage is an attractive shrub native to desert washes and foothill slopes in the deserts of the American Southwest. This species has potential for desert gardens. The shrub is a member of the mint family. The nutlet fruit are borne in inflated pods. The pods can be collected by hand.

Seeds of Bladder Sage germinate without pretreatment.

Salvia—Sage

The genus *Salvia* contains over 500 species in the mint family. Many of these species are herbs, but many semi-woody or woody species are included. Most species have the fragrant foliage associated with the mint flavor. Many species have very attractive flowers borne in whorls on an erect spikate inflorescence. The fruits are nutlets that can be stripped from the plants when mature.

Little information is available on germination of seeds of these species. Generally, a cold-moist stratification period is required. In the case of Creeping Sage (*S. sonomensis*), a soak in gibberellic acid can be substituted for cold-moist stratification.

Sambucus—Elderberry

The elderberries include about 20 species native to temperate and subtropical regions of both hemispheres. The fruits of most species are used by wildlife as well as man.

The fruit is a berrylike drupe containing 3–5 one-seeded nutlets. Elderberry fruits can be collected by stripping or cutting the clusters from the branches.

The fruits may be dried or macerated and the seeds extracted.

Elderberry seeds are difficult to germinate because of their dormant embryo and hard seed coats. A 10–15-minute soak in H_2SO_4 followed by two months' cold-moist stratification at 34°–40°F is suggested.

Sapindus drummondii—Western Soapberry

Western Soapberry grows from Missouri to southern Arizona and northern Mexico. The glossy yellow fruit and long, pinnate leaves make it attractive for environmental plantings.

The fruit, a globular drupe, usually contains a single, dark brown, hard-coated seed. Fruits can be collected any time during late fall or winter by hand-picking. Seed extraction is facilitated by sprinkling the fruits twice daily with water until the pulp softens. The seeds can be extracted by maceration and flotation.

Germination can be improved by treatment with H_2SO_4 for 2–3 hours followed by 90 days' stratification at 35°–45°F.

Satureja douglasii—Yerba Buena

Yerba Buena is a semi-woody member of the mint family. The species is widely distributed in the mountains of far-western North America. It should be considered for use in revegetation of harsh sites.

The fruit is a nutlet that requires no pretreatment for germination.

Shepherdia—Buffaloberry

Shepherdia includes three shrubs native to North America. They provide cover and food for wildlife and are used in shelterbelt plantings.

The drupe-like fruits develop during the summer. The fruits contain achenes enveloped in a fleshy perianth. Fruits are ripe when they turn yellow or red. They may be gathered by stripping or flailing them from the bushes. Heavy gloves are required with Silver Buffaloberry (*S. argentea*) to avoid injury from the thorns. Seeds can be separated by running the fruits through a macerator with water.

Both embryo dormancy and hard-seededness occur in seeds of *Shepherdia*. Acid scarification and cold stratification are recommended.

Simmondsia chinensis—Jojoba

Jojoba is a grayish-green shrub that occurs in southern California, Arizona, and Sonora and Baja, California. Jojoba seeds contain about 50% liquid wax, made up of esters of long-chain alcohols and fatty acids. The oil has been suggested as a replacement for sperm-whale oil.

To avoid seed loss, the fruits should be picked green. Dried fruits can be broken up in a macerator or hammer mill.

Jojoba seeds germinate without pretreatment.

Spiraea betulifolia var. lucida—Birchleaf Spirea

Birchleaf Spirea grows on a wide range of forest sites, from sea-level to subalpine. The showy, flat-topped inflorescences of white flowers make it a desirable species for environmental plantings.

Seeds are borne in a follicle. Little is known about seed collection and germination. Seeds germinate best at low incubation temperatures. Fall sowing has resulted in germination and seedling establishment in the spring after snow-melt.

Staphylea bolanderi—Bladdernut

Erect shrub with opposite three-foliate leaves. Found on the lower slopes of the Sierra Nevada Mountains of California. This is a potential ornamental, specimen, and/or revegetation species.

Fruit is a bladdery capsule. Seeds are smooth and light brown in color. The seeds require acid scarification for germination.

Styrax officinalis—Snowdrop Bush

Styrax consist of a genera of about 100 species in the styrax family. Snowdrop Bush occurs in upper chaparral communities in the western foothills of the California mountains. This is another potentially ornamental, specimen, and/or revegetation species for situations where it is adapted.

The fruit is a capsule containing minute black seeds. The seeds require two months' stratification for germination.

Suaeda—Seep Weed

Suaeda is a genus of about 50 species in the goosefoot family. They have wide distribution, with most species inhabiting saline/alkaline soils. Often these are the only plants that will grow in these extreme situations.

The fruit is a utricle enclosed in the calyx. Seeds are often quite small.

Symphoricarpos—Snowberry

Snowberries occur in North America and Asia. The snowberries have been used in wildlife plantings to provide food and cover. Because of their attractive fruit, the snowberries are desirable species for ornamental plantings.

The fruit is a berrylike drupe ranging in color from white to red, pink, or bluish black. Each fruit contains two nutlets. Ripe fruits can be collected any time during the fall and winter by stripping or flailing onto drop cloths. Seeds can be extracted by macerating the fruits in winter.

Nutlets of *Symphoricarpos* are difficult to germinate because of hard seed coats and immature embryos. Warm stratification for 3–4 months has been used to soften the endocarp. A subsequent period of cold at 41°F for 4–6 months is necessary to induce germination.

Syringa—Lilac

The lilacs include about 25 species of deciduous shrubs native to Asia and southeastern Europe. They are grown primarily for ornament because of their large, showy panicles of flowers, which are often fragrant.

The fruit is a capsule that ripens in late summer and fall. Each capsule contains four thin, flat seeds. The ripe capsules may be picked from the shrubs by hand. They can be run through a hammer mill and then an air-screen.

Apparently, dormancy is variable among species. Cold stratification at 34°–41°F for 30–90 days has been used to enhance germination.

Tamarix pentandra—Five-Stamen Tamarisk

The natural range of Five-stamen Tamarisk is from southeastern Europe to central Asia. It has escaped from cultivation in the United States, and it now grows along major river drainages throughout most of the western United States. Because tamarisk is a heavy user of water, attempts have been made to control the species.

Flowering occurs from March through September. A succession of small capsular fruits ripen and split, releasing minute seeds. The fruits can be collected by hand and sown as seeds. The seeds require no pretreatment.

Tetradymia—Horsebrush

Horsebrush species are low-growing shrubs with dense wool on the stems of most species. They occur in warm and cold desert-plant communities of the Intermountain Area. Members of the sunflower family, the fruit is an achene which often has a dense pappus of whitish bristles. Many of these species are poisonous to animals.

The seeds of the Horsebrush species are very dormant, and germination procedures have not been developed.

Ulex europaeus—Common Gorse

Gorse is a spiny, dense, deciduous shrub. Native to central and western Europe, it has become naturalized in coastal areas of North America, where it is a terrible pest. A member of the legume family, gorse has a pod for a fruit.

The seeds require scarification before germination can occur. Either acid scarification or hot-water treatments have been used.

Vaccinium—Blueberry

The blueberries include about 130 species of deciduous and evergreen shrubs. They are found from the Arctic Circle to high mountains in the tropics.

Blueberries are easily collected by hand-picking the ripe berries. After collection, the berries should be chilled at 50°F for several days to facilitate seed extraction with a macerator.

Dormancy varies among species. The seeds of some require no pretreatment while others require cold-moist stratification.

Viburnum—Viburnum

There are about 120 species of *Viburnum* occurring in the Northern Hemisphere. They are widely used as ornamentals because they have attractive foliage, and showy flowers and fruits.

The fruit is a one-seeded drupe. Much of the wildlife and ornamental value in viburnums is due to persistence of their fruits. The fruits may be hand-picked when ripe. Whole fruits can be dried and used as seed, or the seeds extracted by using a macerator.

Seeds of viburnums are difficult to germinate. Most species have embryo dormancy, and some have hard seed coats. Northern species need warm stratification followed by cold stratification. Southern species require only cold stratification.

Vitis—Grape

About 50 species are widely distributed in temperate parts of the Northern Hemisphere. The species are woody vines climbing by tendrils. The fruit is an ovoid berry.

Ripe fruits can be stripped from the vines by hand or shaken onto drop cloths. Seed can be extracted with a macerator. Seeds of most species require cool-moist stratification for an extended period.

Yucca—Yucca

There are about 30 species of *Yucca* native to North America and the West Indies. Most are long-lived species growing in the arid southwestern United States and Mexico. This is a very important species for desert gardens and for revegetation.

The fruit is a dehiscent capsule containing 100 to 150 seeds. Since the capsules are dehiscent, fruits should be collected just before the capsule opens. Seeds are easily extracted when the capsules are dry.

Pregermination is apparently not needed, but germination is speeded up by cold-moist stratification for some species.

Zanthoxylum—Prickly Ash

Species of *Zanthoxylum* are widely distributed in the eastern United States and Canada. In some areas, they provide food and cover for wildlife.

Fruits are globose, single-seeded capsules. Capsules may be hand-picked from trees. The capsules open, discharging the seed after drying.

Zanthoxylum seeds exhibit strong dormancy. Scarification with H_2SO_4 for two hours is required, followed by cold-moist stratification for 120 days at 41°F.

Zygophyllum fabago var. brachycarpum—Bean-Caper

Bean-caper belongs to the caltrop family, which includes Creosote Bush. *Zygophyllum* includes about 70 species found in the Old World. Bean-caper has been introduced to the American Southwest.

The fruit is an angled capsule. The seeds are gray-brown in color. The seeds contain a water-soluble germination inhibitor that must be leached away before germination can occur.

SUGGESTED ADDITIONAL READING

Acacia—Acacia
Germination:

Everitt, J. H. 1983. Seed germination characteristics of two woody legumes (retama and twisted acacia) from south Texas. *J. Range Manage.* 36:411–414.

Everitt, J. H. 1983. Seed germination characteristics of three woody plant species from south Texas. *J. Range Manage.* 36:246–249.

Acamotopappus sphaerocephalus—Goldenhead
Germination:

Kay, B. L., C. M. Ross, and W. L. Graves. 1977. *Goldenhead.* Mojave Revegetation Notes No. 15. Agronomy and Range Science, University of California, Davis, CA. 3 pp.

Amelanchier—Serviceberry
Germination:

Brinkman, K. A. 1977. *Amelanchier*, pp. 212–215. Seeds of Woody Plants in the United States. Agric. Handbook 450. Forest Service, U.S. Dept. Agric., Washington, DC. Hereafter cited as Handbook 450.

McLean, A. 1967. Germination of Forest Range Species from Southern British Columbia. *J. Range Manage.* 20:321–322.

Adenostoma—Chamise
Germination: Emery, D. 1964. Seed Propagation of Native California Plants. Leaflet of Santa Barbara Botanical Garden 1(10):80–96.

Ailanthus altissima (Mill.) Sumble—Tree of Heaven
Germination: Little, S. 1974. *Ailanthus altissima* (Mill.) Sumble, pp. 201–202. Handbook 450.

Ambrosia dumosa (Gray) Payne—Burrowbrush
Germination and Nursery Practices: Kay, B. L., G. M. Ross, and W. L. Graves. 1977. Burrowbrush. Mojave Revegetation Notes No. 1. Agronomy and Range Science Dept., University of California, Davis, CA. 6 pp.

Amorpha—Amorpha, False Indigo
Germination: Brinkman, K. A. 1974. *Amorpha*, pp. 216–219. Handbook 450.

Aralia—Aralia
Germination: Blum, B. M. 1974. *Aralia*, pp. 220–221. Handbook 450.

Arctostaphylos—Manzanita
Germination: Berg, A. R. 1974. *Arctostaphylos*, pp. 228–231. Handbook 450.

Nursery Practices: Giersbach, J. 1937. Germination and Seedling Production of *Arctostaphylos uva-ursi*. Contrib. Boyce Thompson Inst. 9:71–78.

Aronia—Chokeberry
Germination: Gill, J. D. and F. L. Pogge. 1974. *Aronia*, pp. 232–234. Handbook 450.

Nursery Practices: Chadwick, L. C. 1935. Practices in propagation by seeds—stratification treatment for many species of woody plants described in fourth article of series. *Am. Nurseryman* 62(12):3–9.

Morrow, E. B., G. M. Darrow, and D. H. Scott. 1954. A quick method of cleaning berryseed for breeders. *Am. Soc. Hortic. Soi. Proc.* 63:265.

Artemisia—Sagebrush, Wormwood
Germination: Deitschman, G. H. 1974. *Artemisia*, pp. 235–237. Handbook 450.

Nursery Practices: Plummer, A. P., D. R. Christensen, and S. B. Monsne. 1968. Restoring big-game range in Utah. Utah Fish & Game Publ. 68–3. 183 pp.

| Taxonomy: | Winward, A. H. 1980. Taxonomy and ecology of sagebrush in Oregon. Bull. 642. Agric. Expt. Sta., Oregon State University. 15 pp. |

Asimina—Pawpaw
| Germination: | Bonner, F. T. and L. K. Halls. 1974. *Asimina,* pp. 238–239. Handbook 450. |

Atriplex—Saltbushes
| Germination: | Ansley, R. J. and R. H. Abernethy. 1984. Seed pretreatments and their effects on field establishment of spring seed Gardner saltbush. *J. Range Manage.* 37:509–513. |

Beadle, N. C. W. 1952. Studies in halophytes. I. The germination of seeds and establishment of the seedlings of five species of *Atriplex* in Australia. *Ecology* 33:49–62.

Foiles, M. W. 1974. *Atriplex,* pp. 240–243. Handbook 450.

Young, J. A., B. L. Kay, H. George, and R. A. Evans. 1980. Germination of three species of *Atriplex. Agronomy Journal* 72:705–709.

| Nursery Practices: | Springfield, H. W. 1970. Germination and establishment of four wing saltbush in the southwest. Res. Paper RM-55. Rocky Mt. Forest and Range Expt. Sta., Forest Service, U.S. Dept. Agric. 48 pp. |

Baccharis—Baccharis
| Germination: | Olson, D. F., Jr. 1974. *Baccharis,* pp. 244–246. Handbook 450. |

Berberis—Barberry
| Germination: | Rudolf, P. O. 1974. *Berberis,* pp. 247–251. Handbook 450. |

Beloperone californica Benth—Chuparosa
| Germination: | Emery, D. 1964. Seed propagation of native California plants. Leaflets of the Santa Barbara Botanic Garden 1(10):80–96. |

Brickellia—Bricklebush
| Germination: | Emery, D. 1964. Seed propagation of native California plants. Santa Barbara Botanical Garden Leaflet 1(10):80–96. |

Bumelia lanuginesa (Michx.) Pers—Gum Bumelia
| Germination: | Bonner, F. T. and R. C. Schmidtling. *Bumelia lanuginesa,* pp. 258–259. Handbook 450. |

Calliandra eriophylla Benth—Fairy Duster
Germination: Emery, D. 1964. Seed propagation of native
                      California plants. Leaflets of the Santa Barbara
                      Botanical Garden 1(10):80–96.

Calycanthus occidentalis H. & A.—Sweet-shrub
Germination: Emery, D. 1964. Seed propagation of native
                      California plants. Leaflets of the Santa Barbara
                      Botanical Garden 1(10)80–96.

Campsis radicaus (L.) Seem—Common Trumpetcreeper
Germination: Bonner, F. T. 1974. *Campsis radicaus,* pp. 260–261.
                      Handbook 450.

Caragana arborescens Lam.—Siberian Peashrub
Germination: Dietz, D. R. and P. E. Slabaugh. 1974. *Caragana*
                      arborescens, pp. 262–263. Handbook 450.

Carpenteria california Torr.—Tree Anemone
Germination: Emery, D. 1964. Seed propagation of native Cali-
                      fornia plants. Leaflets of the Santa Barbara
                      Botanical Garden 1(10):80–96.

                      Neal, D. L. 1974. *Carpenteria california,* pp. 76–81.
                      Handbook 450.

Cassia armata Wats—Armed Senna
Germination and   Kay, B. L., C. M. Ross, and W. L. Graves. 1977.
Nursery Practices: Armed senna (*Cassia armata*). Mojave Revegeta-
                      tion Notes No. 2. Agronomy and Range Science,
                      University of California, Davis, CA. 4 pp.

Cassiope mertensiana (Bong.) G. Don.—White Heather
Germination: Nichols, G. E. 1934. The influence of exposure to
                      winter temperatures upon seed germination in
                      various native American plants. *Ecology* 15:364–
                      373.

Castanopsis—Chinkapin
Germination: Hubbard, R. L. 1974. *Castanopsis,* pp. 276–277.
                      Handbook 450.

Ceanothus—Ceanothus
Germination: Bratkowski, H. 1973. Pregermination treatments
                      for redstem ceanothus seeds. Research paper
                      PNW 156. Pacific Northwest Forest and Range
                      Expt. Sta., Forest Service, U.S. Dept. Agric.,
                      Portland, OR.

Emery, D. 1964. Seed propagation of native California plants. Leaflet of Santa Barbara Botanical Garden 1(10):90–91.

Reed, M. J. 1974. *Ceanothus,* pp. 284–290. Handbook 450.

Celastrus scandens L.—American Bittersweet
Germination: Wendel, G. W. 1974. *Celastrus scandens,* pp. 295–297. Handbook 450.

Cephalanthus occidentalis L.—Common Buttonbush
Germination: Bonner, F. T. 1974. *Cephalanthus occidentalis,* pp. 301–302. Handbook 450.

Ceratoides—Winterfat
Germination: Booth, D. T. 1984. Threshing damage to radicle apex affects geotropic response of winterfat. *J. Range Manage.* 37:222–225.

Detari, H. M., J. F. Balliette, J. A. Young, and R. A. Evans. Temperature profiles for germination of two species of winterfat. *J. Range Manage.* 37:218–222.

Springfield, H. W. 1974. *Eurotia lanata,* pp. 398–400. Handbook 450. Seeding Practices: Booth, D. T. and G. E. Schuman. 1983. Seedbed ecology of winterfat: fruits versus threshed seeds. *J. Range Manage.* 36:387–390.

Cercis—Redbud
Germination: Afanasiev, M. 1944. A study of dormancy and germination of seeds of *Cercis canadensis. Jour. Agr. Res.* 69:405–420.

Roy, D. F. 1974. *Cercis,* pp. 305–308. Handbook 450.

Cercocarpus—Mountain Mahogany
Germination: Deitschman, G. H., K. R. Jorgensen, and A. P. Plummer. 1974. *Cercocarpus,* pp. 309–316. Handbook 450.

Liacus, L. G. and C. E. Nord. 1961. Curlleaf cercocarpus seed dormancy yields to acid and thiourea. *J. Range Manage.* 14:317–320. This study reports unnecessary treatments that confuse enhancement procedures.

Young, J. A., R. A. Evans, and D. L. Neal. 1978.

Treatment of curlleaf cercocarpus seeds to enhance germination. *J. Wildlife Manage.* 42:614–620.

Nursery Practices: Plummer, A. P., D. R. Christensen, and S. B. Monsen. 1968. Restoring big-game range in Utah. Utah Division of Fish & Game Publ. 68–3. Salt Lake City, UT. 183 pp.

Cereus giganteus Engelm.—Saguaro
Germination: Alcorn, S. M. and S. C. Martin. 1974. *Saguaro,* pp. 313–314. Handbook 450.

Chamaebatia foliolosa Benth—Mountain Misery
Germination: Emery, D. 1964. Seed propagation of native California plants. Leaflet of Santa Barbara Botanical Garden 1(10)80–96.

Chamaebatiaria millefolium (Torr.) Maxin—Fernbush
Germination: Emery, D. 1964. Seed propagation of native California plants. Leaflet of Santa Barbara Botanical Garden 1(10)80–96.

Cercidium floridum Benth.—Palo Verde
Germination: Emery, D. 1964. Seed propagation of native California plants. Leaflet of Santa Barbara Botanical Garden 1(10)80–96.

Chilopsis linearis (Cav.)—Sweet Desertwillow
Germination: Magill, A. W. 1974. *Chilopsis linearis,* pp. 321–322. Handbook 450.

Chrysothamnus—Rabbitbrush
Germination: Deitschman, G. H., K. R. Jorgensen, and A. P. Plummer. 1974. *Chrysothamnus,* pp. 326–328. Handbook 450.

Clematis—Clematis
Germination: Rudolf, P. O. 1974. *Clematis,* pp. 331–334. Handbook 450.

Colubrina california Jtn.
Germination: Pammel, L. H. and C. M. King. 1928. Germination studies of some trees and shrubs. *Proc. Iowa Acad. Sci.* 36:201–211.

Colutea arborescens L.—Common Bladder-Senna
Germination: Rudolf, P. O. 1974. *Colutea arborescens,* p. 335. Handbook 450.

Comarostaphylis diversifolia (Parry) Greene—Summer Holly

| Germination: | Emery, D. 1964. Seed propagation of native California plants. Leaflet of Santa Barbara Botanical Garden 1(10)80–96. |

Cotoneaster—Cotoneaster

| Germination: | Slabaugh, P. E. 1974. *Cotoneaster,* pp. 349–352. Handbook 450. |

Cowania mexicana var. *stansburiana* (Torr.) Jepson—Cliffrose

| Germination: | Young, J. A. and R. A. Evans. 1981. Germination of seeds of antelope bitterbrush, desert bitterbrush and cliffrose. ARR-W-17. Science and Education Administration, USDA, Oakland, CA. 39 pp. |

Cytisus scoparius (L.) LK.—Scotchbroom

| Germination: | Gill, J. D. and F. L. Pogge. 1974. *Cytisus scoparius* (L.) LK., pp. 370–371. Handbook 450. |

Dalea—Indigobush

| Germination: | Emery, D. 1964. Seed propagation of native California plants. Leaflet of Santa Barbara Botanical Garden 1(10):80–96. |

Dendromecon rigida Benth.—Stiff Bushpoppy

| Germination: | Emery, D. 1964. Seed propagation of native California plants. Leaflet of Santa Barbara Botanical Garden 1(10)80–96. |

Neal, D. L. 1974. *Dendromecon rigida* Benth., p. 372. Handbook 450.

Encelia virginenesis A. Nels. ssp *actoni* (Elmer) Keck—Virgin River Encelia

| Germination and Nursery Practices: | Emery, D. 1964. Seed propagation of native California plants. Leaflet of Santa Barbara Botanical Garden 1(10):80–96. |

Kay, B. L., C. M. Ross, and W. L. Graves. 1977. Virgin River Encelia. Mojave Revegetation Notes No. 4. Agronomy and Range Science, University of California, Davis, CA. 4 pp.

Ephedra—Ephedra

| Germination: | Kay, B. L., C. M. Ross, W. L. Graves, and C. R. Brown. 1977. Gray ephedra and green ephedra. Mojave Revegetation Notes No. 19. Agronomy and Range Science, University of California, Davis, CA. 7 pp. |

Young, J. A., R. A. Evans, and B. L. Kay. 1977. Ephedra seed germination. *Agronomy Journal* 69:209–211.

Epigaea repens L.—Trailing Arbutus
Germination: Blum, B. M. and A. Krochmal. 1974. *Epigaea repens* L., pp. 380–381. Handbook 450.

Eriogonum—Buckwheat
Germination: Kay, B. L., C. M. Ross, and W. L. Graves. 1977. California buckwheat. Mojave Revegetation Notes No. 5. Agronomy and Range Science, University of California, Davis, CA. 4 pp.

Ratliff, R. D. 1974. *Eriogonum fasciculatum* Benth., pp. 382–383. Handbook 450.

Euonymus occidentalis Nutt. extorr.—Burning Bush
Germination: Emery, D. 1964. Seed propagation of native California plants. Leaflet of Santa Barbara Botanical Garden 1(10):80–96.

Fallugia paradoxa (Don) Endl.—Apache Plume
Germination: Deitschman, G. H., K. R. Jorgensen, and A. P. Plummer. 1974. *Fallugia paradoxa* (Don) Endl., pp. 406–408. Handbook 450.

Fouquieria splendens Engelm.—Ocotillo
Germination: Went, F. W. 1984. Ecology of desert plants. I. Observations on germination in the Joshua Tree National Monument. *Ecology* 29:242–253.

Fremontodendron—Fremontia
Germination: Nord, E. C. 1974. *Fremontodendron,* pp. 417–419. Handbook 450.

Galvezia speciosa (Nutt.) Gray—Bush Snapdragon
Germination: Emery, D. 1964. Seed propagation of native California plants. Leaflet of Santa Barbara Botanical Garden 1(10):80–96.

Garrya—Silktassel
Germination: Reynolds, H. G. and R. R. Alexander. 1974. *Garrya,* pp. 420–421. Handbook 450.

Gautheria—Salal
Germination: Emery, D. 1964. Seed propagation of native California plants. Leaflet of Santa Barbara Botanical Garden 1(10):80–96.

Gaylussacia baccata (Waugh.) K. Koch.—Black Huckleberry
Germination: Bonner, F. T. and L. K. Halls. 1974. *Gaylussacia baccata* (Waugh.) K. Koch., pp. 427–428. Handbook 450.

Grayia—Hopsage
Germination: Smith, J. G. *Grayia*, pp. 434–436. Handbook 450.

                      Wood, M. K., R. W. Knight, and J. A. Young. 1976.
                      Splny hopsage germination. *J. Range Manage.*
                      29:53–56.

Hippophae rhamnoides L.—Common Sea Buckthorn
Germination: Slabaugh, P. E. 1974. *Hippophae rhamnoides* L., pp.
                      446–447. Handbook 450.

Holodiscus discolor (Pursh) Maxim—Oceanspray
Germination: King, J. E. 1974. The effect of various treatments to
                      induce germination of seeds of some plants
                      valuable for soil conservation and wildlife. MS
                      thesis. University of Idaho, College of Forestry,
                      Moscow, ID. 97 pp.

                      Stickney, P. F. 1974. *Holodiscus discolor* (Pursh)
                      Maxim, pp. 448–449. Handbook 450.

Hymenoclea salsola T. & G.—White Burrowbush
Germination and   Kay, B. L., C. M. Ross, and W. L. Graves. 1977.
Nursery Practices: Whitebush. Mojave Revegetation Notes No. 7.
                      Agronomy and Range Science, University of Cali-
                      fornia, Davis, CA. 5 pp.

Hyptis emoryi Torr.—Desert Lavender
Germination: Went, F. W. 1948. Ecology of desert plants. I.
                      Observations on germination in the Joshua Tree
                      National Monument. *Ecology* 29:242–253.

Isomeris arborea Nutt.—Bladderpod
Germination: Kay, B. L., C. M. Ross, and W. L. Graves. 1977.
                      Bladderpod. Mojave Revegetation Notes No. 8.
                      Agronomy and Range Science Dept., University of
                      California, Davis, CA. 4 pp.

Kalmia latifolia L.—Mountain Laurel
Germination: Olson, D. F., Jr. and R. L. Barnes. 1974. *Kalmia
                      latifolia* L., pp. 470–471. Handbook 450.

Kochia—Kochia
Germination: Waller, S. S., C. M. Britton, D. K. Schmidt, J.
                      Stubbendieck, and F. A. Sneva. 1983. Germina-
                      tion characteristics of two varieties of *Kochia
                      Prostrata* (L.) Schrad. *J. Range Manage.* 36:242–245.

                      Young, J. A., R. A. Stevens, and R. L. Everett. 1981.
                      Germination of *Kochia prostrata* seeds. *Agronomy J.*
                      73:957–961.

Larrea tridentata Vail—Creosote Bush
Germination: Barbour, M. G. 1968. Germination requirements of the desert shrub Larrea divaricata. Ecology 49:915–923.

Kay, B. L., C. M. Ross, and W. L. Graves. 1977. Creosote Bush. Mojave Revegetation Notes No. 9. Agronomy and Range Science, University of California, Davis, CA. 10 pp.

Martin, S. C. 1974. *Larrea tridentata* Vail, pp. 486–487. Handbook 450.

Lavatera assurgentiflora Kell.—Malva Rose
Germination: Emery, D. 1964. Seed propagation of native California plants. Leaflet of Santa Barbara Botanical Garden 1(10):80–96.

Ledum glandulosum Nutt.—Labrador Tea
Germination: Emery, D. 1964. Seed propagation of native California plants. Leaflet of Santa Barbara Botanical Garden 1(10):80–96.

Myrica—Wax Myrtle
Germination: Emery, D. 1964. Seed propagation of native California plants. Leaflet of Santa Barbara Botanical Garden 1(10):80–96.

Fordham, A. J. 1973. Dormancy in seeds of temperate zone woody plants. Proc. International Plant Propagators Society 23:266–275.

Nemopanthus mucronatus (L.) Trel.—Mountain Holly
Germination: Schopmeyer, C. S. 1974. *Nemopanthus mucronathus* (L.) Trel., p. 553. Handbook 450.

Olney a tesota Gray—Desert Ironwood
Germination: Emery, D. 1964. Seed propagation of native California plants. Leaflet of Santa Barbara Botanical Garden 1(10):80–96.

Osmaronia cerasiformis (Torr. and Gray) Green.—Osoberry
Germination: Dimock, E. J. and W. I. Stein. 1974. *Osmaronia*, pp. 561–563. Handbook 450.

Parkinsonia—Retama
Germination: Everitt, J. H. 1983. Seed germination characteristics of two woody legumes (retama and twisted acacia) from south Texas. *J. Range Manage.* 36:411–414.

Parthenocissus—Creeper
Germination: Gill, J. D. and F. L. Pogge. 1974. *Parthenocissus,* pp. 568–571. Handbook 450.

Peraphyllum ramosissimum Nutt.—Squawapple
Germination: Smith, J. G. 1974. *Peraphyllou ramosissimum* Nutt., pp. 576–577. Handbook 450.

Peucephyllum schottii (Gray) Gray—Pygmy Cedar
Germination: Emery, D. 1964. Seed propagation of native California plants. Leaflet of Santa Barbara Botanical Garden 1(10):80–96.

Philadelphus lewisii Pursh.—Lewis Mock Orange
Germination: Stickney, P. F. 1974. *Philadelphus lewisii* Pursh., pp. 580–581. Handbook 450.

Photinia arbutifolia Lindl.—Christmasberry
Germination: Magill, A. W. 1974. *Photinia arbutifolia* Lindl., pp. 582–583. Handbook 450.

Phyllodoce—Mountain Heather
Germination: Emery, D. 1964. Seed propagation of native California plants. Leaflet of Santa Barbara Botanical Garden 1(10):80–96.

Physocarpus—Ninebark
Germination: Everitt, J. H. 1984. Seed germination characteristics of three woody plant species from south Texas. *J. Range Manage.* 36:246–249.

                      Gill, J. D. and F. L. Pogge. 1974. *Physocarpus,* pp. 584–585. Handbook 450.

Purshia—Bitterbrush
Germination: For a literature review and comparison of germination enhancement treatments, see Young, J. A. and R. A. Evans. 1981. Germination of seeds of antelope bitterbrush, desert bitterbrush, and cliffrose. ARR-W-17. Science and Education Administration, U.S.D.A., Oakland, CA. 39 pp.

Rhododendron—Rhododendron
Germination: Olson, D. F., Jr. 1974. *Rhododendron,* pp. 709–712. Handbook 450.

Rhus—Sumac
Germination: Brinkman, K. A. 1974. *Rhus,* pp. 715–719. Handbook 450.

Ribes—Currant, Gooseberry
Germination: Pfister, R. D. 1974. *Ribes,* pp. 720–727. Handbook 450.

Rosa—Rose
Germination: Gill, J. D. and F. L. Pogge. 1974. *Rosa,* pp. 732–737. Handbook 450.

Steuart, R. N. and P. Semeniuk. 1965. The effect of temperature with afterripening requirement and compensating temperature on germination of seed of five species of *Rosa. American Jour. Botany* 52:755–760.

Rubus—Blackberry, Raspberry
Germination: Brinkman, K. A. 1974. *Rubus,* pp. 738–743. Handbook 450.

Heit, C. E. and G. L. Slate. 1950. Treatment of blackberry seed to secure first year germination. *Proc. American Society Horticulture Sci.* 55:297–301.

Salvia—Sage
Germination: Nord, E. C., L. E. Gunter, and S. A. Graham. 1971. Gibberellic acid breaks dormancy and hastens germination of creeping sage. PSW-259. Research Note. Pacific Southwest Forest and Range Expt. Sta., Forest Service, U.S.D.A. 3 pp.

Sambucus—Elderberry
Germination: Brinkman, K. A. 1974. *Sambucus,* pp. 754–757. Handbook 450.

Sapindus drummondii Hock. & Arn.—Western Soapberry
Germination: Read, R. A. 1974. *Sapindus drummondii* Hock. & Arn., pp. 758–759. Handbook 450.

Satureja douglasii (Benth.) Brig.—Yerba Buena
Germination: Emery, D. 1964. Seed propagation of native California plants. Leaflet of Santa Barbara Botanical Garden 1(10):80–96.

Shepherdia—Buffaloberry
Germination: Thilenius, J. F., K. E. Evans, and E. C. Garrett. 1974. *Shepherdia,* pp. 771–773. Handbook 450.

Simmondsia chinensis (Link) C. K. Schneid—Jojoba
Germination: Nord, E. C. and A. Kodish. 1974. *Simmondsia chinensis* (Link) C. K. Schneid, pp. 774–776. Handbook 450.

Spiraea betulifolia var. lucida (Dougl.) C. L. Hitche.—Spiraea
Germination: Stickney, P. F. 1974. *Spiraea betulifolia* var *lucida*
                        (Dougl.) C. L. Hitche., p. 785. Handbook 450.

Staphylea bolauderi Gray—Bladderpod
Germination: Emery, D. 1964. Seed propagation of native
                        California plants. Leaflet of Santa Barbara
                        Botanical Garden 1(10):80–96.

Styrax officinalis L.—Snowdrop Bush
Germination: Emery, D. 1964. Seed propagation of native
                        California plants. Leaflet of Santa Barbara
                        Botanical Garden 1(10):80–96.

Symphoricarpos—Snowberry
Germination: Evans, K. E. 1974. *Symphoricarpos*, pp. 787–790.
                        Handbook 450.

Syringa—Lilac
Germination: Rudolf, P. O. and P. E. Slabaugh. 1974. *Syringa,* pp.
                        791–793. Handbook 450.

Tamarix pentandra Pall.—Five-Stamen Tamarisk
Germination: Horton, J. S., F. C. Mounts, and J. M. Kraft. 1960.
                        Seed germination and seedling establishment of
                        phreatophyte species. Paper 48. Rocky Mountain
                        Forest and Range Expt. Sta., Forest Service,
                        U.S.D.A. 26 pp.

Ulex europaeus L.—Common Gorse
Germination: Rudolf, P. O. 1974. *Ulex europaeus* L., p. 828.
                        Handbook 450.

Vaccinium—Blueberry
Germination: Crossely, J. A. 1974. *Vaccinium,* pp. 840–843.
                        Handbook 450.

Viburnum—Viburnum
Germination: Barton, L. V. 1958. Germination and seedling
                        production of Viburnum. *Plant Propag. Soc.* 8:126–
                        135.

                        Fordham, A. J. 1973. Dormancy in seeds of
                        temperate zone woody plants. *Proc. Inter. Plant
                        Propagators Society* 23:266–275.

                        Gill, J. D. and F. L. Pogge. 1974. *Viburnum,* pp. 844–
                        850. Handbook 450.

Vitus—Grape
Germination: Mamarau, P., J. Ivanov, and K. Katerov. 1959. The effect of presowing treatment of vine seeds on germination. *Hortic. Abstr.* 29(2222).

Yucca—Yucca
Germination: Arnott, J. H. 1962. The seed, germination, and seedlings of *Yucca.* University California Publ. Botany 35(1):164. University of California Press, Berkeley, CA.

Kay, B. L., C. R. Brown, and W. L. Graves. 1977. Joshua Tree. Mojave Revegetation Notes No. 16. Agronomy and Range Science, University of California, Davis, CA. 4 pp.

Zauthoxylum—Prickly Ash
Germination: Bonner, F. T. 1974. *Zauthoxylum,* pp. 859–861. Handbook 450.

Zygophyllum fabago L. var *brachycarpum* Boiss—Bean Caper
Germination: Keller, D. 1955. Germination regulation mechanisms in some desert seeds. *Bull. Res. Council Israel* 4(4):379–387.

Lerner, H. R., A. M. Mayer, and M. Evenari. 1959. The nature of the germination inhibitors present in dispersal units of *Zygophyllum* and *trigonella. Physiologia Plantarum* 12:245–250.

CHAPTER 11

Germination of Herbaceous Species

The available information on germination of herbaceous species makes it necessary to change from a genus format to a family basis as was used for trees and shrubs. There is another quantum drop in information when we turn to the question of the germination of herbaceous species.

We, therefore, must combine all herbaceous families except for the grass family in this section.

Alismataceae—Water-Plantain Family
A family of mostly aquatic or marsh herbs. About 10 genera mostly of Northern Hemisphere.

Sagittaria—Arrowhead
Mostly perennial aquatic or marsh herbs. Pistils are crowded on a large globose receptacle, forming flat achenes.

Sagittaria latifolia
Wappato or tule-potato is a highly variable species widely distributed in marsh habitats. Seeds require 5–7 months' storage in cold water before even moderate germination will occur (Muenscher 1936).[1]

---

[1]Author and date refer to literature suggested for additional reading at end of this section. No citation indicates information from authors' unpublished research.

Amaranthaceae—Amaranth Family

A large family of about 40 genera and 500 species. Often weedy plants; but a few are grown for ornamental purposes, and several species are grown as pot herbs or in Latin America for their edible seed.

Amaranthus—Pig Weed

Pig weeds are annual, usually coarse herbs. Fruit is a one-seed utricle. The seed is compressed, smooth, consisting of the embryo coiled into a ring around the endosperm. Despite their often dark-colored seed coats, these species usually require light for germination. Best germination at relatively warm temperatures (Schonbeck and Egley 1980).

Amaryllidaceae—Amaryllis or Onion Family

Perennial herbs, with bulbs, corms, or rhizomes. Family consist of 90 genera and about 1,200 species. Many species are valuable horticulturally and economically.

Allium—Onion

This is a huge genus of about 500 species. Seeds of most domestic onions do not require pretreatment. Seeds of native species may require cold-moist stratification. Fruit is a capsule. Seeds are relatively small with wrinkled seed coats.

Apocynaceae—Dogbane Family

Perennial herbs, shrubs, or vines that contain milky juice. Large family of about 150 genera and 1,000 species. Species are widely distributed in warmer regions. Many such as Oleander and Frangipani are important ornamental species.

Apocynum—Dogbane

Perennial herbs with horizontal rootstocks and upright, branching stems with tough fibers. The fruits are follicles and the seeds covered with comose hairs. Despite the fact that some species are serious weeds, there is not appreciable germination information in literature.

Araceae—Arum Family

A large family of about 100 genera and 1,500 species generally found in tropical environments. Many species cultivated as ornamentals.

Arisaema—Jack-in-the-Pulpit

Seeds of A. atrorubens require cool-moist stratification for 60 days (Lincoln 1983). Following pretreatment, seeds should be incubated at 60°–80°F with light. Germination may require a month to be completed.

Asclepiadaceae—Milkweed Family

Perennial herbs, vines, or shrubs with milky juice. Large family of about 200 genera and 2,500 species. Species widely distributed, but most frequently in warmer regions.

Asclepias—Milkweed

Perennial herbs or shrubs from deepseated roots. Fruit a follicle. Seeds covered with long silky hairs. Seeds of Butterfly Milkweed (*A. tuberosa*) require prechilling for 21 days at 35°–41°F prior to germination at moderate temperatures (AOSA 1970). Seeds of Swamp Milkweed (*A. incarnata*) germinate best after cool-moist stratification for five days followed by incubation at 60°F night alternating with 80°F days (Lincoln 1983).

Morrenia—Milkweed Vine

Seeds of Milkweed Vine (*M. odorata*) have highest germination at incubation temperatures from 75°–85°F with a 12-hour photo period (Singh and Achhireddy 1984).

Boraginaceae—Borage Family

Family characterized by perfect flowers in one-sided, scorpioid cymes or racemes. Almost 2,000 species of worldwide distribution, but especially abundant in western United States.

Cryptantha

Annual or perennial herbaceous or suffruticose plants. The fruits are nutlets. *C. intermedia* is a widely distributed species whose nutlets require no pretreatment for germination (Mirov and Kraebel 1939).

Cactaceae—Cactus Family

A large family of the dry tropics and subtropics. Composed of perennial herbaceous or woody succulent plants, with columnar globose or flattened stems.

Opuntia—Prickly Pear

Scarification of seeds of *O. edwardsii*, *O. discata*, and *O. lindheimeri* in concentrated sulfuric acid for 30–60 minutes increased germination (Potter et al. 1984). Seeds of these species that were passed through the digestive tracts of cattle also exhibited enhanced germination compared to non-scarified seeds.

Capparidaceae—Caper Family

Family of herbs or shrubs with ill-smelling foliage and watery sap. About 35 genera and 450 species mostly of warmer regions.

Cleome—Rocky Mountain Bee Plant

Erect glabrous to sparsely pubesent annuals. About 75 species mostly in tropical areas of Americas and Africa. Spiderflower (*C. gigantea*) is grown as an ornamental. Germination of seeds is quite difficult. Light and KNO_3 enrichment required along with warm temperatures. Seeds very sensitive to cold temperatures (AOSA 1970).

Caryophyllaceae—Pink Family
Annual or perennial usually with opposite leaves. About 75 genera and 1,500 species most abundant in temperate and cooler regions, many grown for their flowers.

Dianthus—Pink
Large genus native to the old world. About 200 species, many of which are grown as ornamentals. Fruit is a capsule. The seeds are compressed laterally. Ornamental species include Sweet Wivelsfield (*D. allwoodi*), Sweet William (*D. barbatus*), Carnation (*D. caryophyllus*), Chinapinks (*D. chinensis*), and Grass Pinks (*D. plumarius*) (AOSA 1970).

Gypsolohilia—Baby's Breath
Herbs with scanty leaves and many small flowers arranged in cymose-paniculate inflorescence. Seeds sensitive to incubation temperatures above 50°F (AOSA 1970).

Silene—Catchfly
Seeds of Sleepy Catchfly (*S. antirrhina*) require light for germination, but germinate at a wide range of incubation temperatures from 60°–80°F (Lincoln 1983).

Chenopodiaceae—Goosefoot Family
We have previously met some of the genera of the goosefoot family in the section on germination of shrubs. The family also contains many herbaceous species.

Chenopodium—Goosefoot
A large, cosmopolitan genus containing many weedy species. The fruit is a utricle with a membranous pericarp free from or adherent to the seed. Seed has embryo partly or entirely around the albumen. Seeds of most species germinate readily. Germination often enhanced by nitrate enrichment.

Salsola—Russian Thistle
Very large genus mostly native to Asia. A few species have been introduced to North America, where they are major weeds. Russian Thistle plants produce large numbers of seeds that are widely distributed when the mature plants tumble. Seeds usually have temperature-related afterripening that requires one to two months to satisfy. Once the afterripening is satisfied the seeds germinate very rapidly at a wide range of temperatures (Young and Evans 1972).

Compositae—Sunflower Family
The largest family of flowering plants, containing possibly 950

genera and 20,000 species. Species are chiefly herbaceous and of world-wide distribution.

Achillea—Yarrow

Perennial aromatic herbs with pinnately dissected leaves. Fruit is an achene. Seeds of most species require light for germination (AOSA 1970). Seeds of Common Yarrow, (*A. millefolium*) gave high germination with alternating moderate incubation temperatures (Robocker 1977).

Ambrosia—Ragweed

Winter chilling in the field removes the primary dormancy in Common Ragweed (*A. artemisiifolia* B.) seeds and allows them to germinate in both light and darkness. A secondary dormancy is induced in the spring, and the seeds can again become dormant (Samimy and Khan 1983b).

Anthemis—Dog Fennel

Annual or perennial herbs with alternate incised dentate to pinnately dissected leaves. Achenes more or less compressed. Seeds may have temperature-related afterripening requirement that requires cold temperatures for germination. Light is required for germination of seeds of most species (AOSA 1970). Ellis and Ilnick (1968) found that small seeds of Corn Chamomile (*A. arvensis*) had higher germination than larger seeds.

Aster—Aster

Summer or fall-flowering herbs that are usually rhizomatous or fibrous-rooted perennials. Genus contains about 250 species centered in North America but widely distributed in temperate regions. Mirov and Kraebel (1939) reported seeds of *A. canescens* germinated without pretreatment.

Baileya—Desert Marigold

Annual or perennial densely woolly herbs. Small genera of 3 or 4 species in southwestern United States and adjacent Mexico. Fruit is an achene. Seeds of *B. pleniradiata* require light for germination (AOSA) 1970).

Balsamorhiza—Arrowleaf Balsamroot

Coarse perennial herb ascending from thick taproot. Flowers in large colorful heads. Fruit is a large achene. Achenes require prolonged cool-moist stratification and then incubation at cold temperatures for germination (Young and Evans 1979).

Brickellia

Annual herbs or perennial woody species. Almost 100 species of Western Hemisphere. No information on herbaceous species, but seeds of woody species germinate without pretreatment (Emery 1964).

Carduus—Italian or Musk Thistle

Large genus native to Eurasia and North Africa. Plants are biennial with conspicuously decurrent, spiny leaves. Seeds of most species germinate rapidly without afterripening requirements (Evans et al. 1979).

Centaurea—Star Thistle

Annual or perennial herbs with large or middle-sized heads of tubular purple, violet, pinkish, white, or yellow flowers. A large genus of about 500 species. This includes many ornamental species and several noxious weeds. Seeds may require clipping of radicle end to obtain rapid germination.

Seeds of Black Knapweed (*C. nigra*) require light for germination. These seeds germinate very rapidly when incubated at 75°–85°F (Lincoln 1983).

Chaenactis

Annual, biennial, or sometimes perennial herbs. About 25 species of the western United States. Seeds of *C. artemisiaefolia* and *C. glabriuscula* do not require pretreatment for germination (Miroy and Kraebel 1939).

Chrysopsis—Golden Aster

Low perennial pubescent herbs. Achenes are flattened and usually have double rows of brownish pappus hairs. Seeds of *C. villosa* require no pretreatment for germination (Mirov and Kraebel 1939).

Cirsium—Thistle

Annual biennial or perennial herbs, spiny, with alternate toothed or more usually pinnatified leaves and solitary to clustered heads. About 200 species mostly of the Northern Hemisphere. Many of the species are serious weeds. Seeds of *Cirsium arvense* require light or GA_3 for germination. Best germination is obtained with light and alternating temperatures of 65°–85°F daily (Wilson 1978). Light is required for marked germination of seeds of Flodman Thistle (*C. flodmanii*), but the requirement for light can be overcome by adding GA_3 to the germination substrate (Wilson and McCarty 1984).

Coreopsis—Tickseed

Annual or perennial herbs with showy heads. Achenes have pappus with barbed teeth. About 100 species of warm and temperate regions. Seeds generally germinate without pretreatment. Seeds of some species such as perennial coreopsis (*C. lanceolata*) require light and KNO_3 enrichment for germination (AOSA 1970).

Cosmos—Cosmos

Annual or perennial herbs with long-peduncled heads. About 25 species native to warmer parts of Americas. Several cultivated for their showy flowers. Seeds require light, or light and KNO_3 enrichment for germination (AOSA 1970).

Crepis—Hawks beard

Annual to perennial herbs. Flowers yellow, producing achenes that narrow toward summit. About 200 species mostly found in Northern Hemisphere. The achenes of several species furnish important food for wildlife, especially birds. Seeds of *C. acuminata* germinate without pretreatment.

Echinops—Globe Thistle

Coarse thistle-like herbs mainly native to southern Europe and southwestern Asia. Seeds require light and good moisture supply for germination (AOSA 1970).

Erechtites—Pilewort

Seeds of *E. hieracifolia* require cool-moist stratification for 60 days. The seeds should be incubated at 70°–85°F. Total germination should occur within two weeks (Lincoln 1983).

Erigeron—Wild Daisy

Annual, biennial, or perennial herbs often with sessile leaves. Largely American genus of 200 species. Seeds require light for germination.

Eriophyllum

Annual or perennial woolly herbs or sub-shrubs. Achenes are 4- or 5-angled and linear. Seeds of Catalina Silver Lace (*E. nevinii*) and Golden Yarrow (*E. confertiflorum*) do not require pretreatment for germination.

Eupatorium—Boneset

Seeds of *E. perfoliatum* germinate when incubated at 70°–85°F in the presence of light (Lincoln 1983). Total germination will probably not exceed 50%.

Galinsoga

Annual herbs with opposite leaves and small cymose heads. A small group of widely distributed weeds. Seeds of Hairy Galinsoga (*G. ciliata*) and Small Flower Galinsoga (*G. parviflora*) germinate without pretreatment (Ivany and Sweet 1973).

Gazania—Gazania

Mostly perennial herbs with leafy stems or leaves crowded in a basal tuff. Achenes villous with delicate scarious toothed scales. No pre-germination treatments required.

Grindelia—Gum Plant

Annual, biennial, or usually perennial herbs often resinous around the flower heads. About 50 species native to North and South America. *G. squarrosa* produces disk and ray achenes that differ in their morphology and requirements for germination. Some achenes of both types require light for germination, but the requirement is less pronounced for disk achenes. Cold-moist stratification enhances germination of both types of achenes (McDonough 1975).

Helenium—Sneezeweed
Annual or perennial herbs, with simple or branching stems. The sneezewood achenes are often ribbed. About 40 species restricted to the New World. Seed of *H. bigelovii* do not require pretreatment for germination (Emery 1964).

Helianthus—Sunflower
Coarse annual or perennial herbs. About 60 species native to North and South America. Flowers in heads, fruit large achene. Seeds of domestic sunflowers germinate without pretreatment. Seeds of many of the native sunflower species are highly dormant. A variety of treatments have been used to enhance germination of dormant sunflower seeds. Cold-moist stratification probably provides the best results.

Olivier and Jain (1978) evaluated the germination of populations of seeds of *H. exilis, H. bolanderi,* and *H. exilis* x *bolanderi* hybrids for response to far red light. Red light promoted germination of *H. bolanderi* but not seeds of *H. exilis.*

Hemizonia—Tarweed
Annual or perennial herbs or shrubs, usually with very glandular and aromatic foliage. Usually fall flowering species. Around 30 species found in California and adjacent Mexico. Seeds of *H. kelloggii* Greene require no pretreatment for germination (Mirov and Kraebel 1939).

Heterotheca—Telegraph Weed
Coarse, erect herbs with yellow flower heads. A small group of about three species found in United States and Mexico. Germination studies of seeds of Camphor Weed (*H. subaxillaris*) have revealed that disk achenes germinate at moderate incubation temperatures while ray achenes are dormant (Hwang and Monaco 1978).

Hieracium—Hawkweed
Perennial herbs often hairy or somewhat glandular. The achenes are oblong to columnar with a series of brownish pappus bristles. More than 700 species mostly found in Europe and South America. Seed usually germinate without pretreatment.

Seeds of Rough Hawkweed (*H. scabrum*) are poor germinators. Incubation of seeds of this species at 60°–80°F produced 25% germination after 2 months. Cool-moist stratification did not enhance germination (Lincoln 1983).

Inula—Elecampane
Coarse glandular or hairy herbs. About 100 species native to Old World. Some species naturalized in United States. Seeds require light for germination (AOSA 1970).

Lactuca—Lettuce

Leafy-stemmed herbs with mostly panicled heads. About 90 species of worldwide distribution. Seeds of garden lettuce (*L. sativa*) require light for germination.

Lapsana—Nipplewort

Seeds of *L. communis* germinate readily at warm incubation temperatures. Cool-moist stratification induces dormancy in seeds of this species (Lincoln 1983).

Layia

Vernal annuals with many flowered heads on naked terminal peduncles. All species occur in California, only two species beyond the state. Seeds have temperature-related afterripening and also require light for germination (AOSA 1970).

Lentoden—Fall Dandelion

Seeds of *L. autumnalis* require light for germination, but germinate readily at warm incubation temperatures of 80°–90°F (Lincoln 1983).

Machaeranthera

Annual, biennial, or perennial herbs from a distinct taproot that is often surmounted by a branching caudex. About 25–30 species of temperate western North America. Temperature-sensitive afterripening. Seeds sensitive to temperatures about 45°F.

Madia—Tarweed

Herbs with very glandular and heavy-scented foliage. Small group of less than 20 species found on west coast of North and South America. Seeds of the widely distributed *M. elegans* require no pretreatment for germination (Mirov and Kraebel 1939).

Onopordum—Scotch Thistle

Tall herbs with erect stems and sinuate or pinnatifid, spiny leaves. About 20 species native to Eurasia. Seeds of Scotch thistle (*O. acanthium*) have complex germination requirements. Germination is enhanced by light, GA_3 and KNO_3 enrichment (Young and Evans 1969).

Sanvitalia—Zinnia

Low, mostly branching annual herbs. About four species native to southwestern United States and adjacent Mexico. Light necessary for germination (AOSA 1970).

Senecio—Groundsel

Very large genus of about 1,000 species. Achenes with white soft bristles. With such a large and diverse group, expect numerous types of germination strategies. Seeds of at least some species require light for germination.

Seeds of golden ragwort (*S. aureus*) require 45 days cool-moist stratification. After stratification the seeds should be incubated at 70°–80°F.

Silybum—Milk Thistle

Stout annual or biennial herbs with large prickly leaves. Flower heads are large with purple flowers. Milk thistle is a naturalized weedy pest in North America. Seeds have temperature-related afterripening requirements. The higher the incubation temperature during germination, the longer the afterripening requirement, up to a maximum of five months. Enrichment of the germination substrate with KNO_3 enhances germination (Young et al. 1978).

Tagetes—Marigold

Annual or sometimes perennial strong-scented herbs. About 20 species from New Mexico and Arizona to Argentina. Seeds require light for germination (AOSA 1970).

Taraxacum—Dandelion

Perennials or biennials with leaves in a basal tuff and heads solitary. A rather large genus where many taxa represent apomictic races. Seeds usually germinate without pretreatment (Mexynski and Cole 1974).

Venidium—Monarch-of-the-Veld

Annual or perennial herb more or less tomentese. Native to South Africa, but some naturalized in United States. Seeds require light for germination (AOSA 1970).

Viguiera—Showy Golden Eye

Herbs with leafy, usually branching stems. About 150 species distributed from southern United States to temperate South America. Seeds of Showy Golden Eye (*V. multiflora*) have a prolonged after ripening of 10–12 months. After afterripening requirements are satisfied, germination is moderate at cool temperatures.

Wyethia—Mule's Ear

Coarse perennial herbs from a thick taproot. Flowers in large heads. Fruit an achene. Cold-moist stratification for 90–120 days required for germination (Young and Evans 1979).

Xanthium—Cocklebur

Coarse weedy annuals with stout branching stems. Fruit is an indurate bur, covered with rigid hooked prickles. Probably only three species. Polymorphic seeds in burs, an upper and lower, which have different germination characteristics. The upper seed requires higher oxygen concentrations than the lower. Oxygen requirements for both seeds reduced at higher temperatures (Mayer and Poljakoff-Mayber 1963).

Convolvulaceae—Morning Glory Family

Annual or perennial herbs, chiefly twining or trailing. Fruit is a globular or plump capsule with few rather large seeds. About 50 genera and 1,000 species found in warmer regions.

Convolvulus—Bindweed

Twining or trailing herbs, usually perennial with alternate leaves. Genus of about 200 species of wide distribution. Some species are useful as ornamentals, but many species are very serious weeds. Field Bindweed (*C. arvensis*) is a noxious weed with largely dormant seeds. The seeds have hard seed coats and embryo dormancy that responds to cold-moist stratification (Jordan and Jordan 1982).

Ipomoea—Morning Glory

Twining or trailing herbs of a large genus of warm regions. The sweet potato (*I. batatas*) belongs to this group. Seeds generally germinate without pretreatment. Seeds of *I. hederacea* and *I. lacunosa* require scarification for germination (Gomes et al. 1978).

Jacquemontia—Smallflower Morning Glory

Freshly harvested seeds of Smallflower Morning Glory (*J. tamnifolia*) germinated best after 25–60 seconds of scarification in an electric scarifier equipped with 100–200-grit emery cloth. Optimum temperature for germination was 77°F (Eastin 1983).

Crassulaceae—Stonecrop Family

Herbaceous or somewhat woody succulent plants. About 25 genera and 900 species, many of which are used as ornamentals. The fruit is a one-celled follicle. The seeds are narrow and usually pointed at both ends.

Sedum—Stonecrop

Herbs or subshrubs with alternate leaves. About 300 species of the north temperate zone and the mountains of the tropics. Seeds of Goldmess Sedum (*S. acre*) require at least 8 hours of light in each 24-hour period for optimum germination (AOSA 1970).

Cruciferae—Mustard Family

A large family of herbs, many of which are of economic value (cabbage, mustard, radish, turnip, rutabaga, cauliflower, stocks, etc.). Fruit is a two-celled capsule.. Seeds are usually small with smooth, hard seed coats.

Alyssum

A. alyssoides is naturalized in the western United States. Seeds germinate without pretreatment.

Arabis—Rock Cress

Biennial or perennial herbs with basal leaves. Large genus native to

Americas and Eurasia. Seeds of many species are highly dormant. Seeds of Rock Cress (*A. alpina*) require light and KNO_3 enrichment for germination (AOSA 1970).

Barbarea—Winter Cress
Seeds of *B. vulgaris* require light for optimum germination (Lincoln 1983). The seeds should be incubated at 75°–85°F.

Brassica—Mustard
Erect branching annual or perennial herbs. About 100 species of Europe, Asia, and North Africa. Naturalized as weeds of fields and waste places. Seeds of many species germinate without pretreatment while others require cold-moist stratification, KNO_3 enrichment, and light for maximum germination (AOSA 1970, Miroy and Kraebel 1939).

Cardaria—Hoary Cress
Perennial rhizomatous herbs that are very serious weeds in the United States. Hoary Cress (*C. draba*) is one of the worst noxious weeds in western North America. Seeds of this species germinate without pretreatment.

Descurainia—Tansy Mustard
Annual or biennial herbs native to Eurasia and the Americas. Seeds of *D. pinnata* germinate without pretreatment (Young et al. 1970).

Erysimum—Wallflower
Annual, biennial, or perennial leafy-stemmed herbs. About 90 species of wide distribution in the temperate zone. Seeds of Horterum Wallflower (*E. allionii*) germinate without pretreatment (AOSA 1970).

Isatis—Dyer's Woad
Dyer's Woad (*I. tinctoria*) is the plant from which the word weed is derived. The seeds of this species readily germinate when threshed from the persistent fruit. The fruit contains an effective, concentrated germination inhibitor which leaches from the fruit and can inhibit the germination of seeds of competing species (Young and Evans 1971).

Lepidium—Peppergrass
Annual or suffrutescent perennial plants. About 130 species widely distributed in the United States and temperate parts of the Old World. Seeds usually germinate without pretreatment.

Nasturtium—Water Cress
Mostly small annual to perennial plants found in wet environments. About 50 species native to the temperate zone. Seed germinate best when exposed to light (AOSA 1970).

Sisymbrium—Tumble Mustard
A large genus native to Europe and Asia with several species introduced to North America as weeds. Seeds of Tumble Mustard (*S.*

altissimum) germinate very rapidly without pretreatment (Young et al. 1970).

Stanleya—Prince's Plume
Annual or perennial herbs with a basal rosette of leaves. Fruit is a linear silique, and the numerous seeds are oblong and marginless. Seeds of Prince's Plume (*S. pinnata*) and Panamint Plume (*S. elata*) germinate without pretreatment (Emery 1964).

Thlaspi—Penny Cress
Low, erect annual or perennial herbs with glabrous leaves. About 60 species in this genus with a wide distribution. Seeds of Field Penny Cress (*T. arvense*) germinate without pretreatment.

Cucurbitaceae—Gourd Family
Annual or perennial herbs, mostly with soft stems. Fruit is an indehiscent pepo (fleshy berry structure with rind and fleshy interior). About 90 genera and 700 species mostly of tropical distribution.

Cucurbita—Gourds
Annual and perennial scandant or trailing herbs with fibrous or tuberous roots. Many species grown for food, ornaments, and containers. Seeds of Calabazilla (*C. foetidissima*) germinate without pretreatment (Mirov and Kraebel 1939). Storage of intact Texas gourd (*C. texana*) pepos for 21 days after harvest increased germination of seeds collected 15–47 days after flowering (Oliver et al. 1983).

Sicyos—Burcucumber
Annual climbing vines that can become weeds in cultivated fields. Seeds of *S. angulatus* required either scarification or cold-moist stratification to enhance germination (Mann and Witt 1981).

Cuscutaceae—Dodder Family
Parasitic plants without chlorophyll, the stems slender, twining, yellow to orange in color and fastened to their host by knobs or hausteria. *Cuscuta* is the single genus in the family, consisting of about 100 species of wide distribution. The fruit is a capsule. Seeds have indurate seed coats that require scarification (Dawson 1965, Hutchison and Ashton 1980).

Cyperaceae—Sedge Family
Grasslikeor rushlike herbs, perennials with rhizomes or annuals with fibrous roots. Culms usually solid and terete or variously angled. Fruit is a triangular or lenticular achene. Considering the size and importance of the family, virtually nothing is known about germination ecology.

Cyperus

Perennial or annual herbs with triangular culms. A large genus of about 600 species found in temperate and tropical regions. This genera includes the serious weeds, Yellow Nut Sedge (*C. esculentus*) and Purple Nut Sedge (*C. rotundus*). Highest germination is obtained for seeds of these species with warm incubation temperatures and light during germination (Thullen and Kelley 1979).

Scirpus—Bulrush

Perennial or rarely annual herbs. A large cosmopolitan genus of about 200 species. *S. robustus* is one of the most important waterfowl feed plants. Seeds of this species are often treated with sodium hypochlorate to enhance germination. Germination is highly dependent on seed source. Some sources will consistently germinate, and others are extremely dormant (George and Young 1977).

Eleocharis—Spikerush

Annual or perennial herbs with rhizomes, stolons, or fibrous roots. Cosmopolitan genus of about 140 species inhabiting bogs, shallow ponds, and salt marshes. Dwarf spikerush (*E. coloradoensis*) is an aquatic plant used for biological displacement of water weeds. Seeds of this species require a precise storage program in cold water to permit germination (Yeo and Dew 1977).

Euphorbiaceae—Spurge Family

Herbs, shrubs, or trees usually with a milky acrid sap. Some species are succulents and cactus-like. Fruit is usually a three-lobed capsule. About 280 genera and 8,000 species.

Eremocarpus—Turkey Mullein

Low, broad, heavy-scented annual with stellate, gray pubescence and longer stinging hairs. Single species *E. setigerus* found in western United States. Seeds germinate without pretreatment.

Ricinus—Castor Bean

Horticultural species *R. communis* naturalized in Southwest. Seeds germinate without pretreatment (AOSA 1970).

Euphorbia—Spurge

Monoecious herbs or shrubs with milky acrid juice. A diverse genus of perhaps 1,000 species of mostly temperate regions. Seeds of *E. marginata* require two months' cold-moist stratification for germination (AOSA 1972). Seeds of Wild Poinsettia (*E. heterophylla*) germinate without pretreatment at alternating temperatures (Bannon et al. 1978).

Fumariaceae—Fumitory Family

Herbs with brittle stems and watery juice. Flowers very irregular. Fruit a capsule.

Dicentra—Bleeding Heart

Perennial herbs, the dissected leaves are largely basal. Fruit a several-seeded capsule. Emery (1964) suggests the seeds of Golden Ear-drops (*D. chrysantha*) germinate without pretreatment, while Mirov and Kraebel (1939) failed to germinate seeds of this species. For seeds of Silver Ear-drops (*D. ochroleuca*), Emery (1964) suggests covering seedbed with 0.25 inches of activated charcoal, placing seeds on top and covering with washed sand.

Geraniaceae—Geranium Family

Annual or perennial herbs. Fruits characterized by persistent styles that coil at maturity. Family represented by 11 genera and about 650 species of temperate and subtropical regions.

Geranium—Cranesbill

Herbs with forking stems and swollen nodes. About 250 species of temperate regions. Seeds of most species germinate without pretreatment (AOSA 1970).

Erodium—Filaree

About 60 species of widespread distribution in temperate regions. Several species introduced in western North America, where they are important forage species. Seed coats are hard. The hilum is ruptured naturally by the action of the persistent fruit as the style coils and uncoils (Young et al. 1975).

Hydrophyllaceae—Waterleaf Family

Herbs or shrubs with opposite or alternate leaves. The fruit is a capsule. The family contains 25 genera and about 300 species mostly found in western North America.

Phacelia—Phacelia

Herbs varying from annual to perennial that usually have pubescent foliage and often are glandular. About 200 species found in western North America, some of which are of horticultural value. Seeds of *P. campanularia*, *P. minor*, and *P. tanacetifolia* require light for germination and should be incubated below 60°F. The addition of KNO_3 may improve germination (AOSA 1970).

Hypericaceae—St. John's Wort Family

Herbs or shrubs with opposite, entire leaves. About 10 genera and 300 species of temperate and warm regions. Fruit is a capsule containing many small seeds.

Hypericum—St. John's Wort

Plant glabrous with yellow flowers and leaves, several nerved from base. About 200 species of the Northerh Hemisphere with some species cultivated. Seed of *H. formosum* germinate without pretreatment (Mirov and Kraebel 1939). Seeds of Spotted St. John's Wort (*H. punctatum*) require light for germination. Incubation temperatures should be 70°–85°F for seeds of this species (Lincoln 1983).

Iridaceae—Iris Family
Perennial herbs, most low with simple or branching stems. Fruit is a few to many-seeded loculicidal capsule. A cosmopolitan family with many plants used for ornamental purposes.

Iris—Iris
Perennial herbs with creeping, more or less tuberous rhizomes or bulb-like base. Perhaps 150 species mostly native to the north temperate zone. Seeds of most species require prolonged cold-moist stratification. For *I. longipetala,* three months' cold-moist stratification is required for seed germination (Emery 1964).

Juncaceae—Rush Family
Perennialor sometimes annual herbs, usually found growing in moist places. Stems frequently form creeping rootstocks. Fruits are loculicidal capsules. Family contains 8 genera and perhaps 300 species.

Juncus—Rush
Perennial herbs with stiff, terete or flat leaf blades. A genus of over 200 species most numerous in temperate zone. Plants usually found in wet meadows or other moist environments. Seeds of *Juncus acutus* germinate without pretreatment (Emery 1964).

Labiatae—Mint Family
Aromatic herbs or shrubs mostly with four-angled stems and opposite or whorled or simple leaves. Fruit consist of four one-seeded nutlets included in the persistent calyx. A family of over 150 genera and 3,000 species, widely distributed in temperate and tropical regions.

Marrubium—Horehound
Marrubium vulgare is a widely distributed weed found on Western rangelands and waste places. Seeds germinate best at very warm incubation temperatures.

Salvia—Sage
Herbs or shrubs usually with extremely aromatic herbage. We have already discussed the germination of shrubby species of *Salvia.* There is apparently great diversity among the species of *Salvia* in their germination or dormancy requirements. Seeds of the noxious weed Mediterranean Sage (*S. aethiopis*) as well as seeds of the ornamental species Blue Bedder (*S. farinacea*) and Scarlet Sage (*S. splendens*) require light for germination (AOSA 1970). Seeds of Creeping Sage (*S. sonomenis*) require three months' cold-moist stratification to obtain

appreciable germination (Emery 1964). Seeds of Pitcher Sage (*S. spathacea*) require no pretreatment for germination (Emery 1964).

Monardella

Annual or perennial herbs with herbage with a pleasant odor. About 20 species of western North America. Seeds of *M. macrantha* and Mustang Mint (*M. lanceolata*) require no pretreatment for germination. However, seeds of the widely distributed and highly variable *M. odoratissima* require no pregermination treatment for fresh seeds, but three months' stratification is recommended for stored seeds (Emery 1964).

Leguminosae—Pea Family

A very large family of herbs. Fruit is usually a legume. A huge group of plants with 450–500 genera and thousands of species. We have already discussed the germination of numerous woody species of legumes.

Astragalus

Annual or perennial herbs of a huge group of species estimated at over 2,000. A few species provide valuable forage, but many species contain toxic alkaloids. Major problem is collecting seeds before they are eaten in the pod by insects. The seeds of many species have hard seed coats that require scarification before germination occurs. The seeds of a few species such as *A. antiselli* will germinate without pretreatment (Emery 1964).

Cassia—Senna

Herbs, shrubs, or trees with even-pinnate leaves and usually yellow flowers. A large genus of warm temperate and tropical regions. The waxy seed coat of Sickle Pod (*C. obtusifolia*) seeds limited germination to about 15% without pretreatment (Creel et al. 1968).

Crotalaria—Rattle Box

Large cosmopolitan genus strongly developed in Asia. Seeds of Showy Crotolaria (*C. spectabilis*) do not imbibe water without scarification or breaking of the strophiote (Egley 1979).

Desmanthus—Bundleflower

Velvet Bundleflower (*D. velatinus*) produces both rough and smooth-coated seeds (Haferkamp et al. 1984). Smooth-coated seeds germinate better than rough seeds without pretreatment. After scarification, rough-coated seeds germinate better than smooth-coated seeds.

Lathyrus—Wild Pea

Annual or mostly perennial herbs with rootstocks or sometimes taproots. About 100 species of Northern Hemisphere and South America. As with many legumes, seeds of *Lathyrus* species often have hard seed coats that require scarification as pregermination treatment.

167

Emery (1964) recommended hot water treatments for seeds of *L. splendens* and *L. sulphureus*. Some species of *Lathyrus* are used as agriculture forage or seed plants and have germination standards established (AOSA 1970).

Lotus—Lotus

Annual or perennial herbs. Perhaps 150 species of all continents, but mostly of Northern Hemisphere. Seeds of many native species of *Lotus* are very dormant, with multiple forms of complex dormancy. Seeds of *L. argophyllus, L. crassifolius, L. grandiflorus, L. purshianus,* and *L. strigosus* require hot water treatment to induce germination (Emery 1964). Do not expect high germination after the hot water treatments. Germination standards exist for some species including Bird's Foot Trefoil (*L. corniculatus*) (AOSA 1970).

Lupinus—Lupine

Annual or perennial herbs with alternate palmately compound leaves. Fruit a pod with seeds that have sunken hilum often surrounded by a thickened ring.

Emery (1964) suggested the fresh seeds of perennial species of *Lupinus* do not require pretreatment, but stored seeds require hot water or acid scarification. The seeds of several species of annual *Lupinus* germinate without pretreatment. In this group fall such species as *L. bicolor, L. densiflorus* and *L. stiversii* (Emery 1964). Seeds of the annual *L. succulentus* require hot water treatment for germination. Seeds of *L. truncatus* require prolonged scarification to obtain any germination.

The seeds of the colorful Russell Hybrid Lupines (*L. polyphyllus*) are slow to germinate but require no pretreatment. Seeds of the Texas Blue Bonnet (*L. subcarnosus*) also do not require pretreatment for germination (AOSA 1970).

Medicago—Medick

Annual or perennial herbs native to Eurasia and Africa. Includes the important forage species, Alfalfa (*M. sativa*). The fruits of this group are twisted into burs. The position of the seeds in the bur often influences germination. Seeds of some selections of some species may require hot water or acid scarification (Young et al. 1970). The Rules for Testing Seed (AOSA 1970) provide standards for germination of several species of *Medicago*, but specify no pretreatments other than removing seeds from the burs.

Melilotus—Sweet Clover

Annual or biennial herbs. Native to Eurasia and Africa. Planted as forage species and widely naturalized as ruderal weeds. Standards of germination provided for most species and cultivars (AOSA 1970).

Sesbania—Sesbania

Herbs or shrubs with numerous leaflets on abruptly pinnate

leaves. Pods contain many narrowly oblong seeds. Seeds of Hemp Sesbania (*S. exaltata*) require scarification for enhanced germination. Freshly harvested seeds of Drummond Rattlebox (*S. drummondii*) germinate best after 2.5 to 4 hours' acid scarification in concentrated sulfuric acid. Optimum incubation temperature for germination is 90°–95°F (Eastin 1984).

Trifolium—Clover

Herbs with mostly palmately trifoliate leaves and adnate stipules. About 300 species abundant in the Americas and Africa. Seeds of many *Trifolium* species have hard seed coats that require scarification for germination. Seeds of clovers that are mechanically threshed and cleaned are less likely to be hard than seeds collected and threshed by hand. Other forms of dormancy exist in *Trifolium* seeds. Germination of seeds of these species has been enhanced with ethylene or carbon dioxide enrichment (Young et al. 1970b). Germination standards exist for many cultivars of the various species of clover (AOSA 1970).

Vicia—Vetch

Herbs, mostly vinelike, with pinnate, usually tendril-bearing leaves. About 130 species of the Northern Hemisphere and South America. Germination standards exist for many forage species of vetch whose seeds generally germinate without pretreatment. Some species require cold-moist stratification for 1 week–10 days.

Liliaceae—Lily Family

Leafy-stemmed or scapose perennial herbs, sometimes somewhat woody or climbing, that arise from bulbs, corms, or root stocks which may become somewhat tuberous. Fruit a capsule or berry, mostly several to many-seeded. A family of 2,000 or more species widely distributed in temperate or subtropical regions. Many lily species are of great importance as food or ornamental plants, or as sources of fiber or drugs.

Lilium—Lily

Perennial herbs with scaly bulbs or scaly rootstocks. A genus of almost 100 species of the Northern Hemisphere. This genus of great horticultural interest. Lilies are usually propagated by rootstocks so there is not a lot of information available for germination of seeds of native lilies. Seeds of the Humboldt Lily (*L. humboldtii*) require 3 months' cold-moist stratification for germination (Emery 1964). Seeds of the horticultural Regal Lily (*L. regale*) germinate without pretreatment (AOSA 1970).

Veratrum—False Hellebore

Tall, stout leafy perennial herbs from thick rootstocks. About 12–14 species native to North Temperate Zone. Plants are poisonous to livestock. Seeds of Corn Lily or Skunk Cabbage (*V. californicum*) require prolonged cold-moist stratification (Williams and Cronin 1968).

Linaceae—Flax Family

Herbs or shrubs with simple leaves. Fruit a capsule that contains compressed shining seeds. About 14 genera and 150 species of wide distribution.

Linum—Flax

Annual or perennial herbs with tough fibrous cortex. Seeds are mucilaginous. About 90 species of warm regions. Fresh seeds of Golden Flax (*L. flavum*) and Flowering Flax (*L. grandiflorum*) require KNO_3 enrichment for germination. AOSA (1970) standards for *L. perenne* require light for germination, but seeds of the subspecies *lewisii* germinate without light.

Loasaceae—Loasa Family

Herbs or somewhat woody plants with rough, often stinging hairs. The fruit is a capsule. About 13 genera and 250 species mainly of tropical and temperate America.

Mentzelia—Blazing Star

Annual to perennial herbs that are freely branching and covered with rigid tenacious barbed hairs. Seeds of *M. laevicaulis* germinate without pretreatment (Mirov and Kraebel 1939).

Loranthaceae—Mistletoe Family

Parasitic plants on woody plants. Fruit a berry with glutinous endocarp. About 20 genera in family with 500 species.

Arceuthobium

Yellowish or brownish plants with fragile, jointed stems. Plants are parasitic on conifers. When mature the fruits discharge viscid seeds with great force, and seeds adhere to adjacent branches. Seeds have brief period of afterripening. Germination can be enhanced with treatment with hydrogen peroxide (Wicker 1964).

Lythraceae—Loosestrife Family

A large family of more than 20 genera and 450 species. Species most abundant in the tropical regions of the Americas. Several species of this family grown as ornamentals.

Lythrum—Loosestrife

Seeds of Purple Loosestrife (*L. salicaria*) require 30 days cool-moist stratification before incubation at 75°–90°F to obtain germination. Seeds should be incubated in light (Lincoln 1983).

Malvaceae—Mallow Family

Herbs, usually, with stellate or branching pubescence and

palmately-ribbed and usually lobed leaves. The fruit is a loculicidal capsule or sometimes berry-like. About 50 genera and 1,000 species of temperate and tropical regions. Family includes many economically important species including cotton.

Abutilon—Indian Mallow

Perennial herbs or semi-shrubs widely grown as ornamentals. Seeds of *A. palmeri* germinate without pretreatment (Emery 1964). Seeds of Velvetleaf (*A. theophrasti*) can be impermeable to water and require scarification (Mulliken and Kust 1970). Soaking seeds of Velvetleaf in hot water for 1 hour reduced hard seeds from 99 to 15%. Immersion in hot water was more effective than dry or humid heat. High temperatures acted within a few minutes to reduce hardseededness. Viability of hard seeds was reduced at temperatures above 160°F.

Sphaeralcea—Globe Mallow

Perennial herbs or suffruticose plants usually closely pubescent. About 60 species found in warmer parts of New World and South Africa. Seeds of *S. ambigua*, Desert-Hollyhock, are reported to germinate without pretreatment (Emery 1964). Seeds of *S. grossulariaefolia* have proven to be exceedingly dormant.

Hibiscus—Rose Mallow

Herbs or shrubs that have many uses for horticultural purposes. A genus of 200 species of warmer regions. Seeds germinate without pretreatment (AOSA 1970).

Onagraceae—Evening Primrose Family

Herbs, shrubs, or trees with simple alternate or opposite leaves. Fruit a capsule, berry or nutlike. About 600 species of wide distribution especially in western North America.

Clarkia

Annual herbs with slender to stoutish stems with epidermis exfoliating below. Fruit is a capsule. About 30 species of temperate western North America and Chile. Seeds of *C. unguiculata* germinate without pretreatment (Mirov and Kraebel 1939).

Oenothera—Evening Primrose

Annual to perennial herbs with alternate or basal leaves. Fruit a membranous to woody capsule. Seed of *O. hookeri* germinate without pretreatment (Mirov and Kraebel 1939).

Orobanchaceae—Broomrape Family

Root-parasitic, rather fleshy herbs, without chlorophyll, having alternate scales in place of leaves. Fruit is a capsule containing many very small seeds.

Orobanche—Broomrape

Usually viscid-pubescent plants with purplish to yellowish flowers. Species are native to obscure habitats in North America, but on worldwide basis they constitute some of the worst weeds of agriculture. Seed germination is apparently triggered by chemicals found on or around roots of host species. For example, see Brown et al. (1952).

Oxalidaceae—Oxalis or Wood Sorrel Family

Annual or perennial herbs or even shrubs or trees. Fruit a capsule or berry-like. About 10 genera and 500 species widely distributed in tropical and temperate regions.

Oxalis—Wood Sorrel

Annual or perennial herb, often with bulbous or tuberous underground parts or creeping root stocks. Fruit a loculicidal capsule that often explodes when mature, widely distributing seeds that form mucilage when moistened. *O. corniculata* is a serious weed pest, especially in greenhouses. Seeds germinate without pretreatment.

Paeoniaceae—Peony Family

Large shrubs or herbs with spirally-arranged estipulate leaves. Fruit consist of 2–5 large follicles, each containing several seeds.

Paeonia—Peony

About 30 species of the Northern Hemisphere, mostly from Asia. Native species to North America are perennial herbs. Many of the species are of horticultural value. Seeds of *P. brownii* require 2–½–3 months' cold-moist stratification (Mirov and Kraebel 1939).

Papaveraceae—Poppy Family

Herbs or shrubs with milky yellow or colorless sap. Fruit a capsule which usually dehisces by apical pores or valves. The seeds are usually produced in large numbers. About 25 genera and 200 species that are widely distributed, but most abundant in western North America.

Eschscholzia—California Poppy

Annual or perennial herbs with colorless juice. Fruit a capsule. About 8–10 species from Columbia River to Mexico. Seeds of *E. californica* germinate without pretreatment (Mirov and Kraebel 1939). Seeds of certain lots may benefit from KNO_3 enrichment (AOSA 1970).

Papaver—Poppy

Annual or perennial herbs with yellowish or milky juice. Fruit is a capsule that is dehiscent by transverse pores under the edge of the sessile disk formed by the united stigmas. The seeds are minute and variously striate and pitted. Seeds of many of the horticultural poppies such as *P. glaucum* or *P. nudicaule* require enrichment from KNO_3 for optimum germination (AOSA 1970). Seeds of *P. orientale* also require light for germination, while seeds of *P. rhoeas* germinate without pretreatment.

Phytolaccaceae—Pokeweed Family
Herbs to trees with several-loculed berry as a fruit. About 17 genera and 100 species of warmer parts of Americas and South Africa.

Phytolacca—Pokeweed
Tall, stout herbs with a 5–12-locular berry as a fruit. *P. americana* is a weed in cultivated areas.

Plantaginaceae—Plaintain Family
Herbs from annuals to perennials with basal longitudinally-ribbed leaves and scapose spikes. Fruit is a capsule with flattened or concave seeds. The seeds often have mucilaginous seed coats.

Plantago—Plantain
Major genus of the family with the characteristics given for the family. Common Plantain, *P. major,* is a highly variable species with complex germination requirements (Chandler 1953). The presence of mucilage on the seed coats of *Plantago* species is thought to permit germination on the surface of seed beds (Evans et al. 1974).

Plumbaginaceae—Leadwort Family
Mostly perennial herbs with basal leaves. Fruit is a utricle or achene. About 10 genera and 300 species usually found growing on saline or alkaline soils. Several species grown as ornamentals.

Armeria—Thrift
Tufted acaulescent, perennial herbs. The fruit is membranous and seed rarely dehisces. Seeds of most species germinate at moderate temperatures without pretreatment (AOSA 1970).

Limonium—Marsh Rosemary
Perennial herbs with broad leaves in a basal tuft. Fruit is membranous and indehiscent. Seeds of *L. sinuatum* germinate without pretreatment. The flower heads should be germinated as a unit (AOSA 1970).

Polemoniaceae—Phlox Family
Family of annual or perennial herbs or more rarely shrubs and vines. Fruit is a capsule. About 18 genera and over 300 species that are most numerous in western North America.

Phlox
Perennial or annual herbs either erect, diffuse or cespitose. Several species used as ornamentals. Seeds of *P. drummondii* germinate without

pretreatment, but germination of freshly harvested seeds may be enhanced by the addition of KNO$_3$ (AOSA 1970).

Polygonaceae—Buckwheat Family

Herbs, shrubs, or climbers, and rarely trees. The fruit of the family is a lenticular or angled achene, sometimes enclosed in the ripening perianth, which may become fleshy. About 30–40 genera and 800 species widely distributed in cold and temperate regions. Family includes *Rheum,* Rhubarb, and *Fagopyrum,* Buckwheat.

Eriogonum—Buckwheat

This is a very large and diverse genus with many native species that may prove to be valuable browse or ornamental species. The fruit is a three-angled, sometimes lenticular achene. Seeds of several species of *Eriogonum* will germinate without pretreatment (Emery 1964). These species include *E. fasciculatum* Benth., *E. arborescens, E. latifolium, E. crocatum. E. cernuum* and *E. giganteum.* Emery (1964) reported *E. umbellatum* required three months' cold-moist stratification to obtain germination. In our experience, seeds of most of the subspecies of *E. umbellatum* will germinate without pretreatment.

Rumex—Dock

Seeds of Curly Dock (*R. crispus*) failed to germinate in the dark but germinated readily in light at 65° through 75°F. A secondary dormancy was induced in Curly Dock seeds by 6 days of dark or light incubation at 90°F, by dark incubation at 60°F or by prolonged soaking under chemically-induced moisture stress (Samimy and Khan 1973).

Pontederiaceae—Pickerel Weed Family

Perennial aquatic or bog plants. Fruit is a capsule or achene with copious endosperm.

Eichhornia—Water Hyacinth

Aquatic weed native to the tropics, but has been introduced to warmer parts of North America. Complicated seed production and germination system outlined by Barton and Hotchkiss (1951).

Portulacaceae—Purslane Family

Annual or perennial herbs more or less succulent. Fruit is a capsule. About 20 genera and 200 species of wide distribution in temperate regions.

Montia

Annuals or perennial herbs with somewhat fleshy leaves. Seeds of *M. perfoliata* require two months' cold-moist stratification for germination.

Portulaca

Fleshy herbs with scattered alternate or partly opposite leaves.

Fruit a globular capsule containing many rounded seeds. Seeds of *P. grandiflora* may require cold-moist stratification at 40°F for 3 weeks followed by incubation at warm temperatures with light (AOSA 1970).

Potamogetonaceae—Pondweed Family
Perennial herbs of fresh water, growing from rhizomes. Fruit drupelike when fresh with weedy endocarp. Family consists of single genus. Many of the species of great importance as food for waterfowl.

Potamogeton—Pondweed
Seeds of *P. foliosus, P. amplifolius, P. pectinatus,* and *P. praelongus* require over-winter storage in 40°F water to obtain germination (Muenscher 1936).

Primulaceae—Primrose Family
Family of scapose or caulescent annual or perennial herbs. Fruit a capsule. About 25 genera and 600 species widely distributed, but most common in Northern Hemisphere.

Primula—Primrose
Perennial herbs mostly low from rhizomes. Light and good moisture supply required for seed germination (AOSA 1970). Germination response can be highly variable.

Dodecatheon—Shooting Stars
Rather low perennial herbs with short rootstocks. Capsule contains many small seeds. No treatment is required for germination of seeds of *D. clevelandii* (Emery 1964).

Lysimachia--Loosestrife
Seeds of Fringed Loosestrife (*L. ciliata*) and Whorled Loosestrife (*L. quadrifolia*) both require light for germination. Seeds of these species should be incubated at warm temperatures of 75°–90°F (Lincoln 1983).

Pyrolaceae—Wintergreen Family
Herbaceous perennials from slender and mostly evergreen simple leaves. Also genera of saprophytes or root-parasites without chlorophyll. Fruit a capsule. This family contains several native species that have great potential for natural gardens, but virtually nothing is known about germination of their seeds. The more interesting genera are: Shin Leaf (*Pyrola*), Pipsissewa (*Chimaphila*), Sugarstick (the saprophytic *Allotropa*), Indian Pipe (*Monotropa*), Pine Drops (*Pterospora*), and Snowplant (*Sarcodes*).

Ranunculaceae—Crowfoot Family
A family of herbaceous species with occasional small shrubs or climbers. Fruit usually an achene, but sometimes a follicle or berry, capsule. A relatively large group of 35 genera and about 1,500 species

mostly of North Temperate Zone.

Aquilegia—Columbine
Perennial herbs from thick caudex. The fruit is a follicle with many black, shining seeds. Seeds of both *A. formosa* and *A. eximia* do not require pretreatment for germination (Emery 1964).

Delphinium—Larkspur
Erect branching herbs, usually perennial from a caudex. About 250 species of the North Temperate Zone including some species that are important horticulturally and some that are major stock-poisoning plants. Seeds of Tall Larkspur (*D. barbeyi*) and Low Larkspur (*D. nelsonii*) require cold-moist stratification from 13–19 weeks for germination (Williams and Cronin 1968).

Thalictrum—Meadow Rue
Erect perennial herbs from a short caudex. Fruit is an achene. Seeds of *T. polycarpum* germinate without pretreatment (Emery 1964). Seeds of *T. dioicum* require 60 days' cool-moist stratification. After stratification the seeds can be incubated at 60°–80°F (Lincoln 1983).

Resedaceae—Mignonette Family
Annual or perennial herbs with watery juice. Fruit is a capsule or a berry with many seeds.

Reseda—Mignonette
Erect or decumbent herbs. Seeds of *R. odorata* require light for germination (AOSA 1970).

Rosaceae—Rose Family
We have covered the many important woody species of this family in a previous section. There are several equally important herbaceous genera.

Sanguisorba—Burnet
Chiefly herbs with unequally pinnate leaves. Fruit an achene enclosed in the four-angled dry, thickened flower tube. Seeds of *S. minor* germinate without pretreatment (AOSA 1970).

Geum—Geum
Perennial herbs with rootstocks. Fruit an achene tipped with elongated style. Seeds of *Geum* species germinate without pretreatment, but are sensitive to drying during testing (AOSA 1970). Seeds of Yellow Avens (*G. aleppecum* var *strictum*) require light for germination. The seeds should be incubated at 60°–80°F. A month is required for total germination of these seeds. Seeds of Purple Avens (*G. rivale*) are poor germinators. Some germination is obtained with incubation at 65°–80°F. Cool-moist stratification did not enhance germination (Lincoln 1983).

Potentilla—Cinquefoil

Perennial, sometimes annual herbs. About 250 species of the Northern Hemisphere. Seeds of Rough Cinquefoil (*P. norvegica*) were sensitive to light quality (Taylorsen 1969). A similar requirement has been noted for seeds of *P. glandulosa,* which also require light. Seeds of Tough Fruited Cinquefoil (*P. recta*) require light for germination, but germinate at a wide range of incubation temperatures (60°–80°F) (Lincoln 1983).

Rubiaceae—Madder Family

Herbaceous or woody plants with opposite entire leaves connected by stipules or in whorls. Ovary splits into indehiscent one-seeded carpels when dry. A large, chiefly tropical family.

Richardia—Pusley

Freshly harvested seeds of Florida Pusley (*R. scabra*) required light for germination (Paul et al. 1976).

Santalaceae—Sandalwood Family

A family of herbs, shrubs, and trees with entire leaves. Fruit a drupe or nut with single seed.

Comandra—Bastard Toad Flax

Small perennial herbs from rootstocks. The fruit is drupelike. Seeds germinate without pretreatment.

Sapindaceae

Cardiospermum

Seeds of Balloonvine (*C. balicacabum*) required three hours' scarification in concentrated H_2SO_4 for optimum germination (Johnson et al. 1979).

Saxifragaceae

Herb or shrubs with opposite or alternate leaves without stipules. Fruit a capsule, follicle, or berry. About 700 species mostly of North Temperate region. We have already discussed some of the woody genera. Not much is known about the germination of seeds of the herbaceous genera.

Heuchera—Alum Root

Perennial herbs mostly with stout caudices or rhizomes. Fruit a capsule with two beaks. Seeds are small, ovoid, and covered with minute spines. Seeds of *H. duranii* germinate without pretreatment (Emery 1964).

Scrophulariaceae—Figwort Family

A family of herbs or occasionally shrubs or trees. Fruit a two-celled, mostly two-valved capsule. A family of 200 genera and 3,000 species of wide distribution. In western North America there are several native species of this family that may prove useful in ornamental plantings.

Antirrhinum—Snapdragon

Genus of erect diffuse annual or perennial herbs. Fruit is a capsule that may split open by pores below the base of the style. Seeds of commercial cultivars of snapdragon require light for germination. Seeds of some hybrids may require cold-moist stratification and KNO$_3$ for maximum prompt germination.

Castilleja—Paint Brush

Herbs or suffrutescent plants that may be partially root-parasites. Fruit is a capsule which contains many seeds. Considering how widespread the occurrence of the *Castilleja* species is and the beauty of their flowers, there is virtually nothing known about the germination of the seeds of this group. It is one of the few plant genera without a listing in the *Bibliography of Seeds* (Barton 1969).

Chelone—Turtlehead

Seeds of *C. glabra* require 60 days' cool-moist stratification. The seeds should be incubated at 60°–80°F with light. These seeds will probably only produce 25% germination (Lincoln 1983).

Digitalis—Foxglove

Erect biennial or perennial herbs with alternate leaves. Fruit is ovoid capsule. *D. purpurea* is naturalized along the Pacific Coast. Foxglove seeds require light for germination (AOSA 1970).

Gerardia—False Foxglove

Seeds of *G. pedicularia* require light for germination. The seeds should be incubated at 60°–80°F (Lincoln 1983).

Linaria—Toad Flax

Mostly annual or perennial herbs with opposite or whorled leaves. Fruit is a capsule that is dehiscent by 4 or more pores or slits below the summit. *L. dalmatica* is serious weed pest. Fresh seeds of *Linaria* may require cold-moist stratification for maximum germination response (AOSA 1970). Seeds of Butter and Eggs (*L. vulgaris*) require light for germination. About 50% germination was obtained with incubation at 60°–80°F for 14 days (Lincoln 1983).

Mimulus—Monkey Flower

Annual or perennial herbs often more or less glandular-pubescent. Fruit a cylindrical capsule. Seeds of *M. bifidus, M. guttatus,* and *M. aurantiacus* germinate without pretreatment (Emery 1964). In addition, Mirov and Kraebel (1939) reported seeds of *M. cardinalis* germinate without pretreatment. *Mimulus* hybrids used for ornamentals require light for seed germination (AOSA 1970).

Seeds of Square-stemmed Monkey Flower (*M. tingens*) must be pretreated with 60 days of cool-moist stratification. After stratification, seeds of this species should be incubated at 60–75°F.

Penstemon—Beard Tongue
Perennial herbs or shrubs with opposite leaves. Fruit a septicidal, cartilaginous capsule. Many species appear suitable for natural gardens. Seeds of the *P. antirrhinoides, P. azureus, P. cordifolius,* and *P. spectabilis* germinate without pretreatment (Emery 1964). Mirov and Kraebel (1939) listed the following *Penstemon* species that will germinate without pretreatment: *P. deustus, P. heterophyllus, P. newberryi, P. palmeri,* and *P. centranthifolius.* We have found the germination of seeds of *P. palmeri* can be greatly enhanced with GA_3. Germination standards exist for *P. barbatus, P. grandiflorus, P. laevigatus,* and *P. hirsutus* (AOSA 1970). The standards for all these species include moderate temperatures for incubation.

Verbascum—Mullein
Biennial or perennial herbs with capsule for fruits. Seeds are very small and pitted. Seeds of Common Mullein (*V. thapsus*) have very complex germination requirements. At certain incubation temperatures light is required (Semenza et al. 1978).

Veronica—Speedwell
Annual or perennial, erect or prostrate herbs with opposite leaves. Capsules are flattened shape containing similarly flattened seeds. Seeds of *V. spicata* L. require light for germination (AOSA 1970).

Solanaceae—Nightshade Family
A family of herbs or shrubs with alternate leaves. The fruit of the family is a capsule or a berry. A large family of 100 genera and 3,000 species.

Solanum—Nightshade
Diverse group of herbs and shrubs, some of which are armed. Fruit is a subglobose berry with several to many seeds. The seeds are flattened. Seeds of *S. douglasii* germinate without pretreatment (Emery 1964). The rules for seed germination require light and KNO_3 for germination of *Solanum* species. Seeds of *S. dulcamara* readily germinate at incubation temperatures from 60°–80°F (Lincoln 1983).

Physalis—Ground Cherry
Seeds of Cutleaf Ground Cherry (*P. angulata*) do not germinate at constant temperatures, but have excellent germination at alternating temperatures (Bell and Oliver 1979).

Sparganiaceae—Bur Reed Family
A family of perennial herbs from creeping rootstocks. The indehiscent, nutlike fruit forms a bur. Family of aquatic habitats with

only one genus.

Sparganium—Bur Reed

The seeds of Bur Reed are enclosed in a thick, hard pericarp. Muenscher (1936) experimented with several species of *Sparganium* and only obtained 9% germination from one species.

Sterculiaceae—Redweed Family

Melochia—Redweed

Scarification of freshly harvested seeds of Redweed (*M. corchorifolia*) for 15–30 seconds in an electric scarifier equipped with 100–120-grit emery cloth resulted in enhanced germination (Eastin 1973).

Tropaeolaceae—Tropaeolum Family

A family of succulent prostrate or vine-like herbs. The fruit consist of three indehiscent one-seeded carpels. Single genus with about 50 species of Latin America.

Tropaeolum—Nasturtium

The Garden Nasturtium (*T. majus*) is sparsely naturalized in the warmer parts of the United States. Seeds germinate without pretreatment.

Typhaceae—Cat Tail Family

Tall perennial herbs from creeping rhizomes. The fruit is the well-known cat tail, which slowly dehisces over winter. Major family of freshwater marsh plants composed of a single genus.

Typha—Cat Tail

Muenscher (1936) obtained about 40% germination of *T. latifolia* seeds stored dry after collection. Reduced oxygen increases germination.

Umbelliferae—Carrot Family

A family of aromatic herbs, commonly with hollow stems and alternate leaves. The dry fruit is often winged or ribbed. A large family of 250 genera and 2,000 species of wide distribution and economic importance. Several species are poisonous. There is not a lot known about the germination of native species of *Umbelliferae*. Seeds of the vegetables Celery (*Apium*) and Caraway (*Carum*) require light for germination, while seeds of Dill (*Anethum*) and Carrot (*Daucus*) germinate in the dark (AOSA 1970).

Pastinaca—Parsnip

The Cultivated Parsnip (*P. sativa*) has escaped and naturalized in the northeastern United States. Seeds of the parsnip should be incubated at 70°–85°F. Lincoln (1983) suggests seeds of this species require light for germination, but AOSA (1970) rules do not require light.

Verbenaceae—Verbain Family
Family of herbs, shrubs, or trees with opposite or whorled leaves. The dry fruit consist of 2–4 bony nutlets.

Verbena—Vervain
Annual or perennial herbs with dry fruit enclosed in calyx. Seeds of most species probably require light for germination (AOSA 1970).

Seeds of Blue Verbain (*V. hastata*) require cool-moist stratification for 30 days. After stratification, seeds can be incubated at from 60°–80°F in the presence of light.

Violaceae—Violet Family
In North America the family is largely represented by herbs. Fruit is a capsule that dehisces by valves. Seeds are large with hard seed coat.

Viola—Violet
Not much germination information is available on native species. Seeds of commercial varieties germinate without pretreatment.

LITERATURE CITED

Awang, M. B. and T. J. Monaco. 1978. Germination, growth, development, and control of camphorweed (*Heterotheca subaxillaris*). *Weed Science* 26:51–57.

Barton, L. V. 1969. *Bibliography of seeds.* Columbia University Press, New York. 858 pp.

Barton, A. W. and J. E. Hotchkiss. 1951. Germination of seeds of *Eichhornia crassipes* Solms. Contribution Boyce Thompson Institute. 17:419–434.

Bell, V. D. and L. R. Oliver. 1979. Germination, control, and competition of cut-leaf groundcherry (*Physalis angulata*) in soybeans (*Glycine max*). *Weed Science* 27:133–137.

Brown, R., A. D. Greenwood, A. W. Johnson, A. R. Langdown, A. G. Cong, and N. Sunderlong. 1952. The Orobanche germination factor. I. Concentration of the factor by counter-correct distribution. *Biochemical Journal* 52:571–574.

Chandler, C. 1953. Germination for some species of *Plantago.* Contribution Boyce Thompson Institute 17:265–271.

Creel, J. M., C. S. Hoveland, and G. A. Buchanan. 1968. Germination, growth and ecology of sicklepod. *Weed Science* 16:396–400.

Dawson, J. H. 1965. Prolonged emergence of field dodder. *Weeds* 13:373–374.

Eastin, E. F. 1983a. Smallflower morning glory (*Vacquecmontia tamnifolia*) germination as influenced by scarification, temperature, and seeding depth. *Weed Science* 31:727–730.

Eastin, E. F. 1983b. Redweed (*Melochia corchorifolia*) germination as influenced by scarification, temperature, and seeding depth. *Weed Science* 31:227–231.

Eastin, E. F. 1984. Drummond rattlebox (*Sesbania drummondii*) germination as influenced by scarification, temperature, and seeding depth. *Weed Science* 32:223–225.

Egley, G. H. 1979. Seedcoat impermeability and germination of showy crotalaria (*Crotalaria spectabilis*) seeds. *Weed Science* 27:355–361.

Ellis, J. F. and R. D. Ilnicki. 1968. Seed dormancy in corn chamomile. *Weed Science* 16:111–113.

Emery, Dara. 1964. Seed propagation of native California plants. Leaflets of the Santa Barbara Botanic Garden 1(10)80–96.

Evans, R. A., J. A. Young, and B. L. Kay. 1974. Germination of winter annual species from a rangeland community treated with paraquat. *Weed Science* 22:185–187.

George, H. and J. A. Young. 1977. Germination of alkali bulrush seeds. *J. Wildlife Manage.* 41:791–793.

Gomes, L. F., J. M. Chandler, and G. E. Vaughan. 1978. Aspects of germination, emergence, and seed production of three *Ipomoea* taxa. *Weed Science* 26:245–248.

Haferkamp, M. R., D. C. Kissock, and R. D. Webster. 1984. Impact of presowing seed treatments, temperatures, and seed coats on germination of velvet bundle flower. *J. Range Manage.* 37:185–188.

Horowitz, M. and R. B. Taylorsen. 1984. Hardseededness and germinability of velvet leaf (*Abutilon theophrasti*) as affected by temperature and moisture. *Weed Science* 32:111–115.

Hutchison, J. M. and F. M. Ashton. 1980. Germination of field dodder (*Cuscuta campestris*). *Weed Science* 28:330–333.

Ivany, J. A. and R. D. Sweet. 1973. Germination, growth, development, and control of galinsoga. *Weed Science* 21:41–45.

Johnston, S. K., D. S. Murray, and J. C. Williams. 1979. Germination and emergence of balloonvine (*Cardiospernum balicacobum*). *Weed Science* 27:73–76.

Johnson, S. K., D. S. Murray, and J. C. Williams. 1979. Germination and emergence of hemp sesbania (*Sesbania exaltata*). *Weed Science* 27:290–293.

Jordan, L. S. and J. L. Jordan. 1982. Effects of prechilling on *Convolvulus arvensis* L. seedcoat and germination. *Annals of Botany* 49:421–423.

Lincoln, W. C., Jr. 1983. Laboratory germination methods of some native herbaceous plant species—preliminary findings. *Newsletter*. Association of Official Seed Analysts 57(2):29–31.

Mann, R. K. and W. W. Witt. 1981. Germination and emergence of burcucumber (*Sicyos angulatus*). *Weed Science* 29:83–86.

Mayer, A. M. and A. Poljakoff-Mayber. 1963. *The germination of seeds*. Pergamon Press, Oxford. 236 pp.

Mezynski, R. R. and D. F. Cole. 1974. Germination of dandelion seed on a thermogradient plate. *Weed Science* 22:506–507.

Mirov, N. T. and C. J. Kraebel. 1939. Collecting and handling seeds of wild plants. Forestry Publ. 5, Civilian Conservation Corps., Washington, D.C. 42 pp.

Muenscher, W. C. 1936. Storage and germination of seeds of aquatic plants. Cornell University Agric. Expt. Sta. Bull. 652. 17 pp.

Mulliken, J. A. and C. A. Kust. 1970. Germination of velvetleaf. *Weed Science* 18:561–564.

Oliver, L. R., S. A. Harrison, and M. McClelland. 1983. Germination of Texas gourd (*Cucurbita texana*) and its control in soybeans (*Clycine max*). *Weed Science* 31:700–706.

Oliveri, A. M. and S. K. Jain. 1978. Effects of temperature and light variations on seed germination in sunflower (*Helianthus*) species. *Weed Science* 26:277–283.

Paul, K. B., J. L. Crayton, and P. K. Bisuas. 1976. Germination behavior of Florida pusley seeds. II. Effects of germination-stimulating chemicals. *Weed Science* 24:349–351.

Potter, R. L., J. L. Petersen, and D. N. Ueckert. 1984. Germination responses of *Opontia* spp. to temperature, scarification, and other seed treatments. *Weed Science* 32:106–110.

Roboacker, W. C. 1977. Germination of seeds of common yarrow (*Achillea millefolium*) and its herbicidal control. *Weed Science* 25:456–459.

Samimy, C. and A. A. Khan. 1983a. Secondary dormancy, growth-regulator effect, and embryo growth potential in curly dock (*Rumex crispus*) seeds. *Weed Science* 31:153–158.

Samimy, C. and A. A. Khan. 1983b. Effect of field application of growth regulators on secondary dormancy of common ragweed (*Ambrosia artemisiifolia*) seeds. *Weed Science* 31:299–303.

Santelman, P. W. and L. Evetts. 1971. Germination and herbicide susceptibility of six pigweed species. *Weed Science* 19:51–54.

Schonbeck, M. W. and G. H. Egley. 1980. Redroot pigweed (*Amaranthus retroflexus*) seed germination responses to afterripening, temperature, ethylene, and some other environmental factors. *Weed Science* 28:543–547.

Semenza, R. J., J. A. Young, and R. A. Evans. 1978. Influence of light and temperature on the germination and seedbed ecology of common mullein (*Verbascum thapsus*). *Weed Science* 26:577–581.

Singh, M. and N. R. Achhireddy. 1984. Germination ecology of milkweed vine (*Morrena odorata*). *Weed Science* 32:781–785.

Taylorson, R. B. 1969. Photocontrol of rough cinquefoil seed germination and its enhancement by temperature manipulation and KNO_3. *Weed Science* 17:144–148.

Thullen, R. J. and P. E. Kelley. 1979. Seed production and germination in *Cyperus esculentus* and *C. rotondus*. *Weed Science* 27:502–505.

Wicker, E. F. 1974. Ecology of dwarf mistletoe seed. Research Paper INT-154. Intermountain Forest and Range Experiment Station, Forest Service, U.S. Dept. Agric., Ogden, UT. 28 pp.

Williams, M. C. and E. H. Cronin. 1968. Dormancy, longevity, and germination of seed of three larkspurs and western false hellebore. *Weed Science* 16:381–384.

Wilson, R. G. and M. K. McCarty. 1984. Germination and seedling and rosette development of Flodman Thistle (*Cirsium flodmanii*). *Weed Science* 32:768–773.

Yeo, R. R. and R. J. Dew. 1977. Germination of seed of dwarf spikerush. *Weed Science* 26:425–431.

Young, J. A. and R. A. Evans. 1971. Germination of dyer's woad. *Weed Science* 15:76–78.

Young, J. A. and R. A. Evans. 1972. Germination and establishment of *Salsola* in relation to seedbed environment. Part I. Temperature, afterripening, and moisture relations of Salsola seeds as determined by laboratory studies. *Agronomy Journal* 64:214–218.

Young, J. A. and R. A. Evans. 1979. Arrowleaf balsam root and mule's ear seed germination. *J. Range Manage.* 32:71–74.

Young, J. A., R. A. Evans, R. O. Gifford, and R. E. Eckert, Jr. 1970.

Germination characteristics of three species of cruciferae. *Weed Science* 18:41–48.

Young, J. A., R. A. Evans, and B. L. Kay. 1970a. Germination characteristics of range legumes. *J. Range Manage.* 23:98–103.

Young, J. A., R. A. Evans, and B. L. Kay. 1975. Dispersal and germination dynamics of broadleaf filaree, *Erodium botrys* (Cav.) Bertol. *Agronomy Journal* 67:54–57.

Young, J. A., B. L. Kay, and R. A. Evans. 1970. Germination of cultivars of Trifolium subterraneaum L. *Agronomy Journal* 62:639–641.

CHAPTER 12

Germination of Seeds of the Grass Family

Graminae, or the grass family, is one of the most important groups of plants in virtually every economic category. The fruit of the members is a caryopsis with a starchy endosperm and usually enclosed at maturity in the lemma and palea. There are about 450 genera and 4,500 species in the family.

The grass family is subdivided into 10 tribes. We will follow the subdivisions of tribes and genera in reviewing the germination characteristic of representative species.

TRIBE 1—FESTUCEAE

Bromus—Bromegrass

Low or rather tall annuals or perennials with closed sheaths, usually flat blades and open or contracted panicles of large spikelets. The bromegrasses include about 100 species of the temperate regions of the world.

The AOSA (1970) standards for germination of *Bromus* species call for light for all species although it is optional for some species.

Emery (1964) reported the seeds of *B. grandis* and *B. laevipes* germinated without pretreatment. Similar findings were reported by Mirov and Kraebel (1939) for *B. catharticus.* We have found that seeds of annual species of bromegrass such as *B. tectorum, B. japonicus,* and *B.*

mollis readily germinate. AOSA (1970) standards require light for seeds of *B. mollis,* a requirement that appears unnecessary.

Dactylis—Orchard Grass

Coarse tufted perennial grasses with flat blades and glomerate panicles. Genus consists of 2 or 3 species of the Old World. Seeds of *D. glomeratea* require cold-moist stratification for 7 days and should receive light during germination (AOSA 1970).

Distichlis—Saltgrass

Desert Saltgrass (*D. spicata* var. *stricta*) is an important forage species of the saline-alkaline basins of the western United States. Seed germination is very limited unless extreme diurnal temperature shifts occur. Highest germination occurred with 50°F night and 105°F day temperature regimes (Cluff et al. 1983).

Eragrostis—Lovegrass

Annual or perennial grasses with loose or dense terminal panicles. Genus includes over 100 species of tropical and temperate regions.

Seeds of *E. curvuala* and *E. trichodes* require light and KNO_3 enrichment for germination. Seeds of *E. trichodes* should also receive cold-moist stratification for 6 weeks before germination.

Seeds of Lehmann Lovegrass (*E. lehmaniana*) and Catalina Boer Lovegrass (*Eragrostis curvula* var. *conferta*) had high germination at incubation temperatures between 75° and 100°F (Martin and Cox 1984).

Festuca—Fescue

Annual or perennial plants with spikelets in narrow or open panicles. The fescue grasses include about 100 species of temperate and cool regions, many of which are important forage species.

The germination of species of *Festuca* is quite variable. Seeds of the widely distributed perennial *F. ovina* are not highly germinable (Young et al. 1981). Seeds of the important native forage species *F. idahoensis* are also not highly germinable (Young et al. 1981).

AOSA (1970) standards have required or optimal requirements for light and KNO_3 for most species of *Festuca.* For fresh seeds of most cultivars, cold-moist stratification for 5 days is recommended before incubation.

Holcus—Velvet Grass

Perennial grasses with flat blades and contracted panicles. Seeds of *H. lanatus* require light for germination (AOSA 1970).

Melica—Melic Grass

Perennial grasses where the culms frequently are bulbous at the base. Genus consists of about 60 species found in the colder parts of the world. Apparently there is very little published information on germination of species of *Melica*.

Phragmites—Reed
Perennial reeds with stout leafy culms up to 36 inches tall, topped by large terminal panicles. Seeds of *P. communis* stored dry have given about 15–20% germination after overwinter dry storage while wet-stored seeds did not germinate (Harris and Marshall 1970).

Poa—Bluegrasses
Annual or perennial grasses with leaves that have boat-shaped tips. A large group of about 150 species in all temperate and cool regions in both hemispheres.

In such a large group of species, you should expect variable germination responses. Seeds of the widely distributed and highly variable *P. secunda* generally germinate without pretreatment (Evans et al. 1977). The seeds of the short-lived perennial *P. bulbosa* require cold-moist stratification for 7 days before incubation with KNO_3 enrichment (AOSA 1970). Most of the species of bluegrass that have germination standards specify KNO_3 and/or light (AOSA 1970).

Puccinellia—Alkali Grass
Low, mostly pale annual or perennial grasses. Genus of about 26 species, mostly of saline or alkaline environments. Seeds of *P. lemmonii* from collections made in native stands are not very germinable. Cold-moist stratification and KNO_3 enrichment may enhance germination.

Vulpia—Annual Fescue
Small spiny or emphemeral annual grasses. These annual grasses were formerly known as species of *Festuca*, hence the common name Annual Fescue. Seeds of *V. octoflora* and *V. microstachys* germinate without pretreatment.

TRIBE 2—HORDEAE

Aegilops—Goatgrass
Annuals with flat blades and awned spikes. An Old World genus of weedy species introduced to North America. Seeds of *A. kotschyi* and *A. longissima* germinated best at relatively cool incubation temperatures (Waisel and Adler 1959). Germination inhibitors are probably located in the glumes.

Agropyron—Wheatgrass
Perennial grasses sometimes tufted and sometimes with creeping rhizomes. About 60 species in the genus, mainly in temperate parts of

the world. Many exotic species of wheat grass have been introduced to western North America to revegetate disturbed range lands.

Seeds of the perennial Blue Bunch Wheatgrass *A. spicatum,* which characterizes many plant communities in the northern Intermountain Area, germinate relatively well without pretreatment (Young et al. 1981). Germination is highly variable among sources.

Seeds of the rhizomatious species *A. smithii* generally have poor germination. Seeds of this species should have KNO_3 enrichment to aid in germination (AOSA 1970).

Elymus—Wild Rye

Annual or perennial grasses with flat or, rarely, convolute blades and cylindric spikes. Genus consists of about 45 species mostly of north temperate regions. Seeds of *E. cinereus* germinate without pretreatment, but seed fill is often very low in material collected from native stands (Young and Evans 1981, Evans and Young 1983). Seeds of *E. canadensis* do require cold-moist stratification for two weeks and light during germination (AOSA 1970). Similar requirements are prescribed for seeds of *E. junceus,* and Emery (1964) considered seeds of *E. glaucus* and *E. triticoides* to give satisfactory germination without pretreatment.

Lolium—Ryegrass

Annual or perennial grasses with flat blades and simple erect stems. Seeds of *L. multiflorum* germinate promptly without pretreatment (Young et al. 1975). AOSA (1970) rules specify KNO_3 enrichment, light, and cold-moist stratification for 3 days before germination testing.

Sitanion—Squirreltail

Erect, short-lived perennial grasses with bristly spikes. Genus consists of about 6 species in the western United States. Seeds of *S. hystrix* germinate without pretreatment (Young and Evans 1977).

Taeniatherum—Medusahead

Annual grass that is a very troublesome weed species on Far Western ranges. Livestock do not prefer the herbiage of Medusahead because of the harsh foliage and barbed awns. The seeds of *T. asperum* have temperature-related afterripening and will not germinate above 50°F for about 3 months after maturity (Young et al. 1968).

TRIBE 3—AVENEAE

Aira—Hairgrass

Rather delicate annual grasses with narrow or open panicles of small spikelets. A genus of about 5 species from the warmer parts of Europe. Seeds of *A. caryophyllea* germinate without pretreatment at a wide range of incubation temperatures (Evans et al. 1974).

Arrhenatherum—Tall Oat Grass
Rather tall perennial grasses with flat blades and narrow panicles. Seeds of A. *elatius* germinated without pretreatment (Plummer 1943). The glumes (hulls) can be removed from Tall Oat Grass seeds to aid in ease of drilling (Schwendiman and Mullen 1944).

Avena—Oat
Annual or perennial grasses, low to quite tall, with narrow to open panicles of rather large spikelets. A huge amount of literature exists on the germination of seeds of A. *fatua*, a very serious weed of cultivated grains. A good review of the factors producing dormancy in A. *fatua* seeds is provided by Richardson (1979).

Danthonia—Oatgrass
Tufted perennial grasses with few flowered open or spikelike panicles or rather large spikelets. Seeds of D. *parryi* gave about 50% germination over a wide range of temperatures (Johnston 1961). According to Toole (1939), seeds of D. *spicata* required acid scarification to obtain substantial germination.

Trisetum
Tufted perennial grasses with flat blades and open, shining panicles. About 65 species of colder and temperate regions.
In a comparison of seeds of T. *spicatum* collected from low and high latitudes, the low-latitude seeds germinated slower, but at higher temperatures. Germination of seeds from high-latitude populations was stimulated by light and stratification (Clebsch and Billings 1976).

TRIBE 4—AGROSTIDAE

Agrostis—Bentgrass
Annual or perennial grasses with mostly flattish leaves and panicles of small spikelets. A genus of 100 species in temperate and colder regions. Seeds of A. *gigantea* and A. *canina* require KNO_3 and light for germination (AOSA 1970). Seeds of A. *stolonifera* and A. *tenuis* also require cold-moist stratification before germination.

Alopecurus—Foxtail
Perennial or annual grasses with flat blades and soft, dense, spikelike panicles. Seeds of A. *pratensis* require light for germination.

Aristida—Triple-awned Grass
Tufted annual or perennial grasses with narrow blades and

panicles. Jackson (1928), in his pioneer studies of the germination of range grasses, tested seeds of *A. purpurea* and *A. longiseta* and obtained from 10–60% germination.

Calamagrostis—Reedgrass
Perennial, usually fairly tall grasses, mostly with creeping rhizomes. A genus of about 100 species of cool and temperate regions. Seeds of *C. canadensis* require cold-moist stratification for 5 days before incubation with KNO_3 and light (AOSA 1970).

Muhlenbergia
Perennial or annual grasses either tufted or with scaly rhizomes. Jackson (1928) tested the germination of seeds of *M. porteri* and *M. arenicola*. He obtained at least 10% germination from seeds of both species.

Oryzopsis—Indian Ricegrass
Mostly slender, tufted perennial grasses. *O. hymenoides* is a very important grass in the drier portions of sagebrush and salt desert rangelands. The seed germination of this species has been studied for 50 years.

Seeds are largely dormant without pretreatment. In nature, the collection of the seeds by rodents, who remove the lemma and palea, may be the major natural stand-renewal process. Seeds must be scarified with H_2SO_4 to obtain germination.

Germination of seeds of Indian ricegrass was enhanced by mechanical scarification, but emergence in the field was only slightly enhanced by this treatment (Zemetra et al. 1983).

Sporobolus—Drop Seed
Annual or perennial grasses with flat or involute leaves, and narrow or spreading panicles. Seeds of *S. cryptandrus* require 5 days' cold-moist stratification before incubation with light and KNO_3 enrichment (Toole 1941). Jackson (1928) tested seeds of *S. airoides* and *S. giganteus*. All the species produce at least 40% seed germination. We have found seeds of most collections of *S. airoides* to be quite germinable.

Stipa—Needlegrasses
Tufted perennial grasses with involute leaves and terminal panicles of one-flowered spikelets. Awns are twisted (geniculate). A large genus including several interesting forage species. Seeds of *S. speciosa* germinate without pretreatment (Young and Evans 1982). Seeds of *S. viridula* require cold-moist stratification for 2 weeks followed by KNO_3 enrichment, but still may be poorly germinable (AOSA 1970). *S. thurberiana* seeds apparently require light for germination, while seeds of *S. comata* have their germination enhanced by acid scarification.

Mirov and Kraebel (1939) indicated that seeds of *S. coronata, S. lettermani,* and *S. pulchra* germinated without pretreatment.

TRIBE 5—ZOYSIEAE

Hilaria
Stiff perennial grasses with solid culms and narrow blades. This is the only representative of this tribe in the far western North America. Small genus of about 5 species of southwestern United States to Central America. Jackson (1928) tested seeds of *H. mutica* and obtained from 10–60% germination depending on the seed lot. Hoover (1939) recommended seeding *H. jamesii* and *H. belangeri* seeds collected from native stands.

TRIBE 6—CHLORIDEAE

Bouteloua—Grama Grass
Annual or perennial grasses with florets crowded in one-sided spikelets. Genus characterizes short-grass rangelands. Seeds of *B. gracilis, B. eriopoda* and *B. curtipendula* were germinated by Jackson (1928) during studies in New Mexico. *B. curtipendula* seeds produced good germination, but germination of the other two species was limited.

Buchloe—Buffalo Grass
Gray-green grass forming a dense sod. Plants are dioecious. The pistillate heads form burs. *B. dactyloides* or buffalo grass is the species. Buffalo grass seeds will remain in a germinable condition for some time, with old seeds germinating better than young seeds (Hoover 1939). Seeds should be prechilled (cold-moist stratification) at 40°F for 6 weeks and receive light and KNO_3 during germination (AOSA 1970).

Chloris—Finger Grass
Annual or tufted perennials with flat or folded scabrous blades and sometimes showy and feathery spikes. Seeds of *C. gayana* require light and KNO_3 for germination.

Cynodon—Bermuda Grass
Creeping perennial grasses with stolons or rhizomes, short leaf blades, and the digitate spikes at the summit of erect culms. Seeds of *C. dactylon* require light and KNO_3 for germination. Germination of Bermuda grass seeds is often quite poor. Ahring and Todd (1978) found that alternating temperatures, KNO_3, and light all enhanced germination of *Cynodon* seeds.

Eleusine—Goose Grass

Annual grasses with 2 to several stout spikes, digitate at the summit of the culms. Seeds of *E. indica* required enrichment with KNO_3 for germination (Toole and Toole 1940).

Phleum—Timothy

Annual or perennial grasses with erect culms, flat blades, and dense cylindrical panicles. Seeds of the Cultivated Timothy (*P. pratense*) require cold-moist stratification at 40°F for 5 days, plus light and KNO_3 during germination (AOSA 1970).

TRIBE 7—PHALARIDAE

Anthoxanthum—Vernal Grass

Sweet-smelling annual or perennial grasses with flat blades and spikelike panicles. Seeds of *A. odoratum* require light for germination (AOSA 1970).

Phalaris

Annual or perennial grasses with many flat blades and dense spikelike panicles. About 20 species in temperate Europe and America.

Germination of seeds of *P. arundinacea* was enhanced by light quality, chilling, and soaking in an oxygen-enriched solution (Landgraff and Junttila 1979).

TRIBE 8—ORYZEAE

This is the tribe of which cultivated rice (*Oryza sativa*) is a member. There are very few native members of the tribe in North America.

TRIBE 9—PANICEAE

Cenchrus—Burgrass or Buffelgrass

Annual or perennial grasses with flat blades and simple racemes of spiny burs. Seeds of *C. ciliaris* have variable germination depending on the seed source. Cultivars of this grass used in Texas have quite dormant seeds that require cold-moist stratification with the burs pressed into well-packed soil for 7 days before germination with light (AOSA 1970). We have tested strains from North Africa that germinate without pretreatment.

Echinochloa—Barnyard Grass

Coarse annual grasses with compressed sheaths; long, flat blades; and terminal panicles of stout, short, densely flowered, one-sided racemes. Seeds of the serious weed pest *E. crusgalli* germinate without pretreatment (AOSA 1970).

Panicium

Annual or perennial grasses of various habitat. A very large genus of both hemispheres.

Seeds of *P. obtusum* germinated best with alternating temperature regimes of 50°–90°F daily (Toole 1940). Germination was enhanced by KNO_3 enrichment. It was necessary to acid-scarify the seeds for 90 minutes before incubation (Hoover 1939). Seeds of *P. dichotomiflorum* were found to be dormant at harvest (Brecke and Duke 1980). The after-ripening lasted for about 6 months. Acid scarification and alternating temperatures enhanced germination.

Paspalum—Knot Grass

Mostly perennial grasses with racemes digitate or racemose of the summit of the culm and branches. Seeds of *P. dilatatum* require light and KNO_3 for germination (AOSA 1970). Seeds of *P. urvillei* require light and KNO_3. For some seed lots, scarification may be required (Mathews 1947).

Pennisetum—Kikuyu Grass

Annual or perennial grasses, often branched, usually with flat blades and dense, spikelike panicles. Seeds of *P. americanum* and *P. purpureum* germinate without precontrol (AOSA 1970).

Setaria—Bristly Foxtail

Annual or perennial grasses with flat leaves and cylindrical spikelike panicles. Seeds of *S. macrostachya* had enhanced germination after acid scarification for 15–30 minutes (Toole and Toole 1940). Highest germination of acid-scarified seed was obtained with incubation at temperature regimes that alternated from 45°–80°F daily. Seeds of *S. lutescens* had variable germination depending on the biotype tested (Norris and Schener 1980). Most sources of seed required a prolonged afterripening period before they would germinate.

TRIBE 10—ANDROPOGONEAE

Andropogon—Bluestem

Rather coarse perennial grasses with solid, leafy stems. About 150 species of warmer regions. Several species of *Andropogon* characterize tall-grass rangelands. Seeds of *A. gerardii* and *A. hallii* require cold-moist stratification for 2 weeks followed by incubation with light and KNO_3 (AOSA 1970).

LITERATURE CITED

Ahring, R. M. and G. W. Todd. 1978. Seed size and germination of hulled and unhulled Bermuda grass seeds. *Agronomy Journal* 70:667–670.

AOSA. 1970. Rules for testing seeds. Proc. of the Association of Official Seed Analysis 60(2):1–116, including all revisions through 1981 *Journal of Seed Technology* 6(7):1–125.

Brecke, B. J. and W. B. Duke. 1980. Dormancy, germination, and emergence characteristics of fall panicum (*Panicum dichotomiflorum*) seed. *Weed Science* 28:683–685.

Clebsch, E. E. and W. D. Billings. 1976. Seed germination and vivipary from a latitudinal series of populations of the arctic-alpine grass *Trisetum spicatum. Arctic and Alpine Res.* 8:255–262.

Cluff, G. J., R. A. Evans, and J. A. Young. 1983. Desert saltgrass seed germination and seedbed ecology. *J. Range Manage.* 34:419–422.

Emery, Dara. 1964. Seed propagation of native California plants. Leaflets of the Santa Barbara Botanic Garden 1(10):81–96.

Evans, R. A. and J. A. Young. 1983. "Magnar Basin wild rye germination in relation to temperature. *J. Range Manage.* 36:395–398.

Evans, R. A., J. A. Young, and B. L. Kay. 1974. Germination of winter annual species from a rangeland community treated with paraquat. *Weed Science* 22:185–187.

Evans, R. A., J. A. Young, and B. A. Roundy. 1977. Seedbed requirements for germination of Sandberg bluegrass. *Agronomy Journal* 69:817–820.

Fulbright, T. E., E. F. Redente, and A. M. Wilson. 1983. Germination requirements of green needlegrass (*Stipa ultidulatin*). *J. Range Manage.* 37:390–394.

Harris, S. W. and W. H. Marshall. 1960. Experimental germination of seed and establishment of seedlings of *Phragmites communis. Ecology* 41:395.

Hover, M. M. 1939. Native and adapted grasses for conservation of soil and moisture in the Great Plains and western states. Farmer Bulletin 1812. U.S. Dept. Agric., Washington, D.C. 43 pp.

196

Jackson, C. V. 1928. Seed germination of some New Mexico range plants. *Botanical Gaz.* 86:270–294.

Landgraff, A. and O. Junttila. 1979. Germination and dormancy of reed canary-grass seeds (*Phalaris arundinacea*). *Physilogia Plantarum* 45:96–101.

Martin, M. H. and J. R. Cox. 1984. Germination profiles of introduced lovegrasses at six constant temperatures. *J. Range Manage.* 37:507–509.

Mathews, A. C. 1947. Observations on methods of increasing the germination of *Panicum anceps* Michx. and *Paspalum notatum* Flugge. *J. American Society Agron.* 39:439–442.

Mirov, N. T. and C. J. Kraebel. 1939. Collecting and handling seeds of wild plants. Forestry Publication #5. Civilian Conservation Corps., Washington, D.C.

Norris, R. F. and C. A. Schoner, Jr. 1980. Yellow foxtail (*Setaria lutescens*) biotype studies: Dormancy and germination. *Weed Science* 28:159–163.

Plummer, A. P. 1943. The germination and early seedling development of twelve range grasses. *J. Amer. Soc. Agronomy* 35:19–34.

Richardson, S. G. 1979. Factors influencing the development of primary dormancy in wild oat seeds. *Can. J. Plant Science* 59:777–784.

Schwendiman, J. L. and L. A. Mullen. 1944. Effect of processing on germinative capacity of seed of tall oat grass, *Arrhenatherum elatius* (L.) Mart. and Koch. *J. Amer. Society Agron.* 36:783–785.

Toole, E. H. 1941. Factors affecting the germination of various dropseed grasses (*Sporoholus spp*). *J. Agr. Res.* 62:691–715.

Toole, E. H. and V. K. Toole. 1940. Germination of seed of goose grass (*Eleusime indica*). *J. Am. Society of Agronomy* 32:320–321.

Toole, V. K. 1940. Germination of seed of vine-mesquite, *Panicum obtusum*, and plains bristle-grass, *Setaria macrostachya*. *J. American Society Agronomy* 32:503–512.

Waisel, J. and Y. Adler. 1959. Germination behavior of *Aegitops Kotsehy*; Boiss. *Can. J. Botany* 37:741–742.

Young, J. A. and R. A. Evans. 1977. Squirrel tail seed germination. *J. Range Manage.* 30:33–36.

Young, J. A. and R. A. Evans. 1981. Germination of Great Basin wildrye seeds collected from native stands. *Agronomy Journal* 73:917–920.

Young, J. A., R. A. Evans, and R. E. Eckert, Jr. 1968. Germination of medusahead in response to temperature and afterripening. *Weed*

Sci. 16:92–95.

Young, J. A., R. A. Evans, and R. E. Eckert, Jr. 1981. Temperature profiles for germination of bluebunch and beardless wheatgrass. *J. Range Manage.* 34:84–89.

Young, J. A., R. A. Evans, and B. L. Kay. 1975. Germination of Italian ryegrass seeds. *Agronomy Journal* 67:386–389.

Young, J. A., R. A. Evans, R. E. Eckert, Jr., and R. D. Ensign. 1981. Germination-temperature profiles for Idaho and sheep fescue and Camby bluegrass. *Agronomy Journal* 73:716–720.

Zemetra, R. S., C. Haustad, and R. L. Cuony. 1983. Reducing seed dormancy in Indian ricegrass (*Oryzepsis hymenoides*). *J. Range Manage.* 36:239–241.

Glossary

Abnormal seedlings. In seed testing, seedlings which do not possess all normal structures required for growth, nor show the capacity for continued development.

Achene. A small, dry fruit containing a single seed which nearly fills the cavity; the seed coat does not adhere to the pericarp.

Adnate. Union or fusion of unlike parts; for example, stamens and petals.

Afterripening. Physiological processes in seeds (or bulbs, tubers, and fruits) after harvest or abscission, which occur prior to and are often necessary for germination or resumption of growth under favorable external conditions.

Alternate. Located singly at different heights, such as leaves.

Anther. Pollen-bearing part of stamen.

Anthesis. Flowering, more precisely, the time of pollination.

Apetalous. Without petals.

Aril. An appendage or outer covering of a seed, common to ovules of euonymous.

Articulate. Jointed.

Axil. Angle formed by plant stem and leaf.

Berry. A fleshy fruit without a stone, usually contains several seeds embedded in pulp.

Bilabiate. Two-lipped (calyx or corolla); for example, the flowers of the mint family.

Bract. A reduced leaf subtending a flower, usually associated with an inflorescence.

Calyx. The whorl of sepals, outer circle of floral envelope.

Campanulate. Bell-shaped.

Capsule. A dry fruit which opens by splitting along the back of individual carpels or along the line where two meet.

Carpel. Ovule-bearing unit of ovary.

Carpelled. Possessing or composed of carpels.

Catkin. An inverted spikelike inflorescence of willow, birch, and oak.

Cleft. Divided nearly to the middle, as with compound leaves.

Cone. A dry strobili bearing seed in some gymnosperms.

Connate. Union or fusion of like parts, such as petals forming a tube.

Corolla. The whorl of petals, inner circle of floral envelope.

Cotyledon. Modified leaf or leaves of the embryo or seedling, which may contain the stored food reserves of the seed. They are formed at the first node or at the upper end of the hypocotyl.

Cruciform. Petals forming a cross.

Dehiscence. Splitting open at maturity as in fruits.

Dioecious. Male and female flowers on different plants.

Dormancy, Embryo. Dormancy as a result of conditions within the embryo itself: inhibiting substances, cotyledon influences, impermeable structures. *Syn.:* Internal Dormancy.

Dormancy, Imposed. Dormancy as a result of some action, treatment, or injury to seeds in the course of collecting, handling, or sowing. *Syn.:* Secondary Dormancy, Induced Dormancy.

Dormancy, Induced. *Syn.:* Dormancy, Imposed; Dormancy, Secondary.

Dormancy, Internal. *Syn.:* Dormancy, Embryo.

Dormancy, Physiological. An embryo dormancy due to physiological conditions which can be overcome by pretreatments other than scarification.

Dormancy, Seed coat. Dormancy as a result of seed coat conditions: impermeability to gases or moisture or mechanical restrictions.

Drupe. Usually a one-seeded fruit with a fleshy outer layer covering a stone.

Embryo. Rudimentary plant within a seed.

Empty seed. A seed testing term for a seed unit which does not contain all tissues essential for germination. This condition can result from insect or disease attack, or incomplete development of the ovule. Intact seed coats devoid of internal tissue are considered "empty seeds" under this concept.

Endocarp. Inner layer of the fruit wall or pericarp.

Endosperm. Storage tissue of the seed.

Epicotyl. Portion of the axis of a plant embryo or seedling stem between the cotyledons and the primary leaves.

Epigeal germination. Germination in which the cotyledons are forced above the ground by the elongation of the hypocotyl.

Epigynous. Borne on or arising from the ovary, inferior ovary.

Evergreen. Plants with persistent leaves, remain green in dormant seasons.

Exocarp. Outer layers of pericarp, the skins on fleshy fruits.

Female gametophyte. Haploid nutrient storage tissue in seeds of gymnosperms. It is often mistakenly called the "endosperm" of seeds of gymnosperms.

Filament. Most commonly, the stalk of the stamen.

Filled seed. A seed with all tissues essential for germination.

Floret. An individual flower.

Flower. The reproductive part of angiosperms.

Follicle. A dry, pod-like fruit which opens along one side only.

Forb. Any herb that is not a grass or grasslike.

Fruit. The ripened ovary.

Funnelform. A flower with a gradually expanding floral tube.

Germination. Resumption of active growth in an embryo which results in its emergence from the seed and development of those structures essential to plant development.

Germination capacity. Proportion of a seed sample that has germinated normally in a specified test period, usually expressed as a percentage. *Syn.:* Germination percentage.

Germination energy. That proportion of germination which has occurred up to the time of peak germination, the time of maximum germination rate, or some preselected point, usually 7 test days. (The critical time of measurement can be chosen by several means.)

Germination percentage. *Syn.:* Germination capacity.

Glabrous. Not hairy, smooth.

Glaucous. Covered with a whitish bloom that rubs off.

Hard seeds. Seeds which remain hard and ungerminated at the end of a prescribed test period because they have not absorbed water because of an impermeable seed coat.

Head. A short, dense inflorescence of sessible flowers, a compressed spike.

Herb. A flowering plant whose stem above ground does not become woody or persistent.

Herbaceous. Not woody, dying back to the ground each year.

Hilum. A scar from the ovular connection.

Hypocotyl. That part of the embryonic axis which is between the cotyledons and the radicle. In seedlings, the juvenile stem which is between the cotyledons and the root system.

Hypogeal germination. Germination in which the cotyledons remain in the seed below the ground while the epicotyl elongates.

Hypogynous. All floral parts are below the ovary, superior ovary.

Imbibition. The mechanism of initial water uptake by seeds. The taking up of fluid by a colloidal system.

Imbricated. Overlapped, in a shingle-like fashion.

Immature embryo. Condition in which a morphologically immatur embryo delays germination.

Imperfect. A flower lacking either male or female sex organs.

Incomplete. A flower lacking one of the perianth whorls.

Indehiscent. Remaining closed.

Inferior. Referring to an ovary which is below the floral envelope.

Inflorescence. The arrangement of flowers on a floral axis.

Inhibition. A restraining or repression of a function of a seed.

Integument. The outer envelope of an ovule, which becomes the seed coat.

Involucre. Whorl of bracts at the base of an inflorescence.

Kernel. Edible portion of a seed embryo or seed storage tissues.

Legume. A dry fruit splitting at maturity along two sutures.

Locule. A chamber or cell of an anther, ovary, or fruit.

Loculicidal. A dry fruit which splits open on the back directly into a locule (see also poricidal and septicidal).

Mesocarp. Middle layer of the fruit or pericarp wall.

Micropyle. Minute opening in the integument of an ovule through which the pollen grain or pollen tube passes to reach the embryo sac. It is usually closed in the mature seed to form a superficial scar.

Monoecious. Male and female flowers on same plant.

Naked stratification. Chilling of seeds without the use of a moisture-holding medium.

Nut. One-seeded, hard, bony-walled, indehiscent fruit.

Opposite. Two at a node, referring to leaves.

Ovary. Ovule bearing part of the pistil.

Ovule. The egg-containing unit of the ovary, which becomes the seed.

Pappus. Sepals reduced to modified hairs or scales and persistent in fruit of compositae.

Perfect. A flower containing both pistil and stamens.

Perianth. A collective term for calyx and corolla.

Pericarp. In angiosperms, a fruit wall which developed from the ovary wall; it may be dry, hard, or fleshy. (*See:* Seed coat.)

Petal. A unit of the corolla.

Petaloid. Petal-like, resembling a petal.

Petiole. The stalk of the leaf.

Petiolule. The stalk of a leaflet in a compound leaf.

Physiological maturity. A general term for the stage in the life cycle of a seed when development is complete and the biochemical components necessary for all physiological processes are active or ready to be activated.

Pistil. The female structure in a flower, a unit of the gynoecium composed of stigma, style, and ovary.

Plumule. Primary bud of a plant embryo situated at the apex of the hypocotyl. Portion of the seedling axis above the cotyledons consisting of leaves and an epicotyl, which elongates to form the primary stem.

Polyembryony. Formation of two or more embryos from a single ovule in a seed.

Prechilling. Cold-moist treatment applied to seeds to overcome dormancy before sowing in soil or germination in the laboratory.

Pretreatment. Any kind of treatment applied to seeds to overcome dormancy and hasten germination.

Purity. Proportion of clean, intact seed of the designated species in a seed lot, usually expressed as a percentage by weight.

Radicle. Portion of the axis of an embryo from which the primary root develops.

Sample, submitted. The sample of seed submitted to a seed testing station.

Sample, working. A reduced seed sample taken from the submitted sample in the laboratory, on which some test of seed quality is made.

Scarification. Disruption of seed coats, usually by mechanical abrasion or by brief chemical treatment in a strong acid, to increase their permeability to water and gases, or to lower their mechanical resistance.

Seed. A matured ovule which contains an embryo and nutritive tissue and is enclosed in protective layers of tissue (seed coat).

Seed coat. Protective outer layer of a seed derived from the integuments of the ovule. When two coats are present, the thick outer coat is the testa and the thin inner coat is the tegmen.

Seed lot. A specified quantity of seed of reasonably uniform quality.

Seed quality. A general term that may refer to the purity, germination capacity, or vigor of a seed lot.

Sepal. A unit of the calyx.

Septate. Partitioned or divided.

Septicidal. A dry fruit which splits into the interior partitions and not directly into a locule.

Septum. A partition or cross-wall.

Sound seed. A seed which contains in viable condition all tissues necessary for germination. *Syn.:* Viable seed.

Spathe. A bract surrounding the flower-cluster.

Stamen. Male organ of the flower, composed of an anther and filament and producing pollen.

Staminode. A sterile stamen.

Stellate. Bearing forked or branched hairs.

Stigma. The apical part of the pistil.

Stippule. One of a pair of usually foliaceous appendages found at the base of the petiole in many plants.

Stratification. Practice of placing seeds in moist medium, often in alternate layers, to hasten afterripening or overcome dormancy. Commonly applied to any technique which keeps seeds in a cold and moist environment.

Strobilus. Male or female fruiting body of the gymnosperms.

Style. The central portion of the pistil, between the stigma and style.

Suture. A line of union, or seam, between two members.

Sympetalous. Having petals joined.

Tegmen. The inner seed coat, usually thin and delicate.

Terrestrial. A land plant, of the ground as opposed to aquatic.

Testa. The outer coat of a seed, usually hard or tough.

Tolerance. A permitted deviation (plus or minus) from a standard. In seed testing, the permitted difference between or among replicated measurements beyond which the measurements must be repeated.

Tree. A perennial plant having a permanent woody, self-supporting main stem or trunk.

Umbel. An inflorescence in which the pedicels of the flowers arise from the same point.

Unisexual. Having only one sex.

Valve. A separable part of a pod, the units or pieces into which a capsule splits.

Vernalization. Treatment of seeds, bulbs, or seedlings with low temperatures (32° to 40°F) to hasten flowering of the subsequent plant.

Viability. The state of being capable of germination and subsequent growth and development of the seedling.

Viable seed. *Syn.:* Sound seed.

Vigor. Those seed properties which determine the potential for rapid, uniform emergence and development of normal seedlings under a wide range of field conditions.

Whorl. Three or more leaves or flowers at a node.

Wing. In seeds, a thin, dry, membranaceous extension.

Xerophyte. A plant of an acid or dry habitat.

Appendix

List of plant material discussed by growth form (trees, shrubs, herbaceous species, and grasses) and scientific names.

TREE SPECIES

Conifers

FAMILY	GENERA	SPECIES
Cupressaceae	*Calocedrus*	*C. decurrens*
	Chamaecyparis	
	Cupressus	*C. macnabiana*
	Juniperus	*J. occidentalis*
Pinaceae	*Abies*	*A. magnifica*
		A. venusta
	Cedrus	
	Larix	
	Picea	*P. breweriana*
	Pinus	*P. edulis*
		P. jeffreyi
		P. lambertiana
		P. monophylla
		P. murrayana
		P. quadrifolia
	Pseudotsuga	*P. macrocarpa*
		P. menziesii

FAMILY	GENERA	SPECIES
Pinaceae	*Thuja*	
	Tsuga	*T. mertensiana*
Taxaceae	*Taxus*	
	Torreya	
Taxodiaceae	*Sequoia*	sempervirens
	Sequoiadendron	giganteum
	Taxodium	distichum
	Torreya	*T. california*

Broadleaf Trees

FAMILY	GENERA	SPECIES
Aceraceae	*Acer*	*A. saccharum*
Anacardiaceae	*Cotinus*	*C. obovatus*
Aquifoliaceae	*Ilex*	
Betulaceae	*Alnus*	
	Betula	
	Corylus	
Bignoniaceae	*Catalpa*	*C. bignonoides*
		C. speciosa
Cornaceae	*Cornus*	
Ebanaceae	*Diospyros*	*D. virginiana*
		D. texana
Elaecagnaceae	*Elaeagnus*	*E. angustifolia*
		E. commutata
Ericaceae	*Arbutus*	*A. menziesii*
Fagaceae	*Castanea*	*C. dentata*
	Fagus	
	Lithocarpus	*L. densiflorus*
	Quercus	*Q. kelloggii*
Hamamelidaceae	*Liquidambar*	*L. styraciflua*
Hippocastanaceae	*Aesculus*	
Juglandaceae	*Carya*	
	Juglans	*J. hindsi*
Lauraceae	*Sassafras*	*S. albidum*
		S. varifolium
	Umbellularia	*U. california*
Leguminosae	*Gleditsia*	
	Gymnocladus	*G. dioicus*
	Robinia	
Magnoliaceae	*Liriodendron*	*L. tulipifera*
	Magnolia	

FAMILY	GENERA	SPECIES
Moraceae	Maclura Morus	M. pomifera
Myricaceae	Myrica	M. california
Myrtaceae	Eucalyptus	
Nyssaceae	Nyssa	
Oleaceae	Fraxinus	
Platanaceae	Platanus	P. occidentalis P. orientalis P. racemosa
Rhamnaceae	Rhamnus	R. alnifolius R. california R. crocea R. purshiana
Rosaceae	Crataegus Malus Prunus Sorbus	
Rutaceae	Ptelea	P. trifoliata
Salicaceae	Populus Salix	
Sapindaceae	Sapium	S. sebiferum
Tiliaceae	Tilia	T. americana
Ulmaceae	Ulmus	

SHRUB SPECIES

FAMILY	GENERA	SPECIES
Acanthaceae	Beloperone	B. californica
Anacardiaceae	Rhus	
Annonaceae	Aisimina	A. parviflora A. trilcha
Aquifoliaceae	Nemopanthus	N. mucronatus
Araliaceae	Aralia	
Berberidaceae	Berberis	
Bignoniaceae	Campsis Chilopsis	C. radicans C. linearis
Buxaceae	Simmondsia	S. chinensis
Cactaceae	Cereus	C. giganteus
Calyxcanthaceae	Calycanthus	C. occidentalis

FAMILY	GENERA	SPECIES
Capparidaceae	*Isomeris*	*I. arborea*
Caprifoliaceae	*Lonicera*	
	Symphoricarpos	
	Viburnum	
Celastraceae	*Celastrus*	*C. scandens*
	Euonymus	*E. occidentalis*
Chenopodiaceae	*Atriplex*	*A. gardneri*
	Ceratoides	*C. lanata*
		C. latens
	Grayia	*G. spinosa*
	Kochia	*K. americana*
		K. prostrata
	Sarcobatus	*S. vermiculatus*
	Suaeda	
Compositae	*Acamotopappus*	*A. sphaerocephalus*
	Ambrosia	*A. dumosa*
	Artemisia	*A. spinescens*
		A. tridentata
	Baccharis	
	Brickellia	*B. californica*
	Chrysothamnus	*C. nauseosus*
		C. viscidiflorus
	Encelia	*E. virginensis*
	Haplopappus	
	Hymenoclea	*H. salsola*
	Lepidospartum	*L. squamatum*
	Peucephyllum	*P. schottii*
	Tetradymia	
Crossosomataceae	*Crossosoma*	*C. californicum*
Cruciferae	*Lepidium*	*L. fremontii*
Elaeagnaceae	*Shepherdia*	*S. argenta*
Ephedraceae	*Ephedra*	*E. nevadenis*
		E. virdis
Ericaceae	*Arctostaphylos*	*A. uva-ursi*
	Cassiope	*C. mertensiana*
	Comarostaphylis	*C. diversifolia*
	Epigaea	*E. repens*
	Gaultheria	
	Gaylussacia	*G. baccata*
	Kalmia	*K. latifolia*
	Ledum	*L. glandulosum*
	Lindera	*L. benzion*
	Phyllodoce	
	Rhododendron	
	Vaccinium	
Fagaceae	*Castanopsis*	*C. chrysophylla*
		C. sempervirens

FAMILY	GENERA	SPECIES
Fouquieriaceae	*Fouquieria*	*F. splendens*
Garryaceae	*Garrya*	
Labiatae	*Hyptis*	*H. emoryi*
	Salazaria	*S. mexicana*
	Salvia	*S. sonomensis*
	Satureja	*S. douglasii*
Leguminosae	*Acacia*	*A. berlandieri*
		A. rigidula
		A. schaffneri
	Amorpha	*A. californica*
		A. canescens
	Calhaudra	*C. eriophylla*
	Caragana	*C. arborescens*
	Cassia	*C. armata*
	Cercidium	*C. floridum*
	Cercis	
	Colutea	*C. arborescens*
	Cytisus	*C. scoparius*
	Dalea	*D. spinosa*
	Lespedeza	
	Olneya	*O. tesota*
	Parkinsonia	*P. aculeata*
	Porlieria	
	Ulex	*U. europaeus*
Liliaceae	*Yucca*	
Malvaceae	*Lavatera*	*L. assurgentiflora*
Menispermaceae	*Menispermum*	*M. canadense*
Myricaceae	*Myrica*	*M. california*
		M. hartwegii
Oleaceae	*Menodora*	*M. scabra*
	Syringa	
Papaveraceae	*Dendromecon*	*D. rigida*
Polygonaceae	*Eriogonum*	
Ranunculaceae	*Clematis*	
Rhamnaceae	*Ceanothus*	
	Colubrina	
	Hippophae	*H. rhamnoides*
Rosaceae	*Adenostoma*	*A. fascicalatum*
		A. sparsifolium
	Amelanchier	*A. laevis*
	Aronia	*A. arbutifolia*
		A. melanocarpa
		A. prunifolia
	Cercocarpus	*C. betuloides*
		C. ledifolius
		C. montanus

FAMILY	GENERA	SPECIES
Rosaceae	Chamebatia	C. foliolosa
	Chamaebatiaria	C. millefolium
	Coleogyne	
	Cotoneaster	
	Cowania	C. mexicana
	Fallugia	F. paradoxa
	Holodiscus	H. discolor
	Lyonothamnus	L. floribundus
	Osmaronia	O. cerasiformis
	Peraphyllum	P. ramosissimum
	Photinia	P. arbutifolia
	Physocarpus	
	Potentilla	P. glandulosa
	Purshia	P. glandulosa
		P. tridentata
	Rosa	
	Rubus	
	Sambucus	
	Spiraea	S. betulifolia
Rubiaceae	Cephalanthus	C. occidentalis
	Mitchella	M. repens
Rutaceae	Zanthoxylum	
Sapindaceae	Sapindus	S. drummondii
Sapotaceae	Bumelia	B. lanuginosa
Saxifragaceae	Carpenteria	C. californica
	Philadelphus	P. lewisii
	Ribes	
Scrophulariaceae	Galvezia	G. speciosa
Simarubaceae	Ailanthus	A. altissima
Solanaceae	Lycium	
Staphyleaceae	Staphylea	S. bolanderi
Sterculiaceae	Fremontia	
Styroceae	Styrax	S. officinalis
Tamaricaceae	Tamarix	T. pentandra
Vitaceae	Parthenocissus	P. quinquefolia
	Vitis	
Zygophyllaceae	Larrea	L. tridentata
Zygophyllaceae	Zygophyllum	Z. fabaso

HERBACEOUS SPECIES

FAMILY	GENERA	SPECIES
Alismatacae	*Sagittaria*	*S. latifolia*
Amaranthaceae	*Amaranthus*	
Amaryllidaceae	*Allium*	
Apocynaceae	*Apocynum*	
Araceae	*Arisaema*	*A. atrorubens*
Asclepiadaceae	*Asclepias*	*A. incarnata*
	A. tuberosa	
	Morrenia	*M. odorata*
Boraginaceae	*Cryptantha*	*C. intermedia*
Cactaceae	*Opontia*	*O. discata*
		O. edwardsii
		O. lindheimeri
Capparidaceae	*Cleome*	*C. gigantea*
Caryophyllaceae	*Dianthus*	*D. allwoodi*
		D. barbatus
		D. caryophyllus
		D. chinensis
		D. plumarius
	Gypsophilia	
	Silene	*S. antirrhina*
Chenopodiaceae	*Chenopodium*	
	Salsola	
Compositae	*Achillea*	*A. millefolium*
	Ambrosia	*A. artemisiifolia*
	Anthemis	*A. arvensis*
	Aster	*A. canescens*
	Baileya	*B. plenitadiata*
	Balsamorhiza	
	Brickellia	
	Carduus	
	Centaurea	*C. nigra*
	Chaenactis	*C. artemisiaefolia*
	C. glabriuscula	
	Chrysopsis	*C. villosa*
	Cirsium	*C. arvense*
		C. flodmanii
	Coreopsis	*C. lanceolata*
	Cosmos	
	Crepis	*C. acuminata*
	Echinops	
	Erechtites	*E. hieracifolia*
	Erigeron	
	Eriophyllum	*E. nevinii*
		E. confertiflorum

FAMILY	GENERA	SPECIES
Compositae	*Eupatorium*	
	Galinsoga	*G. ciliata*
		G. parviflora
	Gazania	
	Grindelia	*G. squarrosa*
	Helenium	*H. bigelovii*
	Helianthus	*H. exilis*
		H. bolanderi
	Hemizonia	*H. kelloggii*
	Heterotheca	*H. subaxillaris*
	Hieracium	*H. scabrum*
	Inula	
	Lactuca	*L. sativa*
	Lapsana	*L. communis*
	Layia	
	Lentoden	*L. autumnalis*
	Machaeranthera	
	Madia	*M. elegans*
	Onopordum	*O. acanthium*
	Sanvitalia	
	Senecio	*S. aureus*
	Silybum	
	Tagetes	
	Taraxacum	
	Venidium	*V. multiflora*
	Viguiera	
	Wyethia	
	Xanthium	
Convolvulaceae	*Convolvulus*	*C. arvensis*
	Ipomoea	*I. batatas*
		I. hederacea
		I. lacunosa
	Jacquemontia	*J. tamnifolia*
Crassulaceae	*Sedum*	*S. acre*
Cruciferae	*Alyssum*	*A. alyssoides*
	Arabis	*A. alpina*
	Barbarea	*B. vulgaris*
	Brassica	
	Cardaria	*C. draba*
	Descurainia	*D. pinnata*
	Erysimum	*E. allionii*
	Isatis	*I. tinctoria*
	Lepidium	
	Nasturtium	
	Sisymbrium	*S. altissimum*
	Stanleya	*S. elata*
		S. pinnata
	Thlaspi	*T. arvense*
Cucurbitaceae	*Cucurbita*	*C. foetidissima*
		C. texana
	Sicyos	*S. angulatus*

FAMILY	GENERA	SPECIES
Cuscutaceae	*Cuscuta*	
Cyperaceae	*Cyperus*	*C. esculentus*
		C. rotundus
	Scirpus	*S. robustus*
	Eleocharis	*E. coloradoensis*
Euphorbiaceae	*Eremocarpus*	*E. setigerus*
	Ricinus	*R. communis*
	Euphorbia	*E. marginata*
		E. heterophylla
Fumariaceae	*Dicentra*	*D. chrysantha*
		D. ochroleuca
Geraniaceae	*Geranium*	
	Erodium	
Hydrophyllaceae	*Phacelia*	*P. campanularia*
		P. minor
		P. tanacetifolia
Hypericaceae	*Hypericum*	*H. formosum*
		H. punctatum
Iridaceae	*Iris*	*I. longipetala*
Juncaceae	*Juncus*	*J. acutus*
Labiatae	*Marrubium*	*M. vulgare*
	Salvia	*S. aethiopis*
		S. farinacea
		S. sonomenis
		S. spathacea
		S. splendens
	Monardella	*M. lanceolata*
		M. macrantha
		M. odoratissima
Leguminosae	*Astragalus*	*A. antiselli*
	Cassia	*C. obtusifolia*
	Crotalaria	*C. spectabilis*
	Desmanthus	*D. velatinus*
	Lathyrus	*L. splendens*
		L. sulphureus
	Lotus	*L. argophyllus*
		L. corniculatus
		L. crassifolius
		L. grandiflorus
		L. purshianus
		L. strigosus
	Lupinus	*L. bicolor*
		L. densiflorus
		L. polyphyllus
		L. stiversii
		L. subcarnosus
		L. succulentus
		L. truncatus

FAMILY	GENERA	SPECIES
Leguminosae	*Medicago*	*M. sativa*
	Melilotus	
	Sesbania	*S. drummondii*
		S. exaltata
	Trifolium	
	Vicia	
Liliaceae	*Lilium*	*L. humboldtii*
		L. regale
	Veratrum	*V. californicum*
Linaceae	*Linum*	*L. flavum*
		L. grandiflorum
		L. perenne
Loasaceae	*Mentzelia*	*M. laevicaulis*
Loranthaceae	*Arceuthobium*	
Lythraceae	*Lythrum*	*L. salicaria*
Malvaceae	*Abutilon*	*A. palmeri*
		A. theophrasti
	Sphaeralcea	*S. ambigua*
		S. grossulariaefolia
	Hibiscus	
Onagraceae	*Clarkia*	*C. unguiculata*
	Oenothera	*O. hookeri*
Orobanchaceae	*Orobanche*	
Oxalidaceae	*Oxalis*	
Paeoniaceae	*Paeonia*	*P. brownii*
Papaveraceae	*Eschscholzia*	*E. californica*
	Papaver	*P. glaucum*
		P. nudicaule
		P. orientale
		P. rhoeas
Phytolaccaceae	*Phytolacca*	*P. americana*
Plantaginaceae	*Plantago*	*P. major*
Plumbaginaceae	*Armeria*	
	Limonium	*L. sinuatum*
Polemoniaceae	*Phlox*	*P. drummondii*
Polygonaceae	*Eriogonum*	*E. arborescens*
		E. cernuum
		E. crocatum
		E. fasciculatum
		E. giganteum
		E. latifolium
		E. umbellatum
	Rumex	*R. crispus*

FAMILY	GENERA	SPECIES
Pontederiaceae	*Eichhornia*	
Portulacaceae	*Montia*	*M. perfoliata*
	Portulaca	*P. grandiflora*
Potamogetonaceae	*Potamogeton*	*P. amplifolius*
		P. foliosus
		P. pectinatus
		P. praelongus
Primulaceae	*Dodecatheon*	*D. clevelandii*
	Lysimachia	*L. ciliata*
		L. quadrifolia
	Primula	
Pyrolaceae	*Allotropa*	
	Chimaphila	
	Monotropa	
	Pterospora	
	Pyrola	
	Sarcodes	
Ranunculaceae	*Aquilegia*	*A. formosa*
		A. eximia
	Delphinium	*D. barbeyi*
		D. nelsonii
	Thalictrum	*T. dioicum*
		T. polycarpum
Resedaceae	*Reseda*	*R. adorata*
Rosaceae	*Geum*	*G. aleppecum*
		G. rivale
	Potentilla	*P. glandulosa*
		P. norvegica
		P. recta
	Sanguisorba	*S. minor*
Rubiaceae	*Richardia*	*R. scabra*
Santalaceae	*Comandra*	
Sapindaceae	*Cardiospermum*	*C. balicacabum*
Saxifragaceae	*Heuchera*	*H. duranii*
Scrophulariaceae	*Antirrhinum*	
	Castilleja	
	Chelone	*C. glabra*
	Digitalis	*D. purpurea*
	Gerardia	*G. pedicularia*
	Linaria	*L. dalmatica*
		L. vulgaris
	Mimulus	*M. aurantiacus*
		M. bifidus
		M. cardinalis
		M. guttatus
		M. tingens

FAMILY	GENERA	SPECIES
Scrophulariaceae	*Penstemon*	*P. antirrhinoides*
		P. azureus
		P. barbatus
		P. centranthifolius
		P. cordifolius
		P. deustus
		P. grandiflorus
		P. heterophyllus
		P. hirsutus
		P. laevigatus
		P. newberryi
		P. palmeri
		P. spectabilis
	Verbascum	
	Veronica	*V. spicata*
Solanaceae	*Physalis*	*P. angulata*
	Solanum	*S. douglasii*
		S. dulcamara
Sparganiaceae	*Sparganium*	
Sterculiaceae	*Melochia*	*M. corchorifolia*
Tropaeolaceae	*Tropaeolum*	
Typhaceae	*Typha*	*T. latifolia*
Umbelliferae	*Pastinaca*	*P. sativa*
Verbenaceae	*Verbena*	*V. hastata*
Violaceae	*Viola*	

GRASS SPECIES

TRIBE	GENERA	SPECIES
Agrostidae	*Agrostis*	*A. canina*
		A. gigantea
		A. stolonifera
		A. tenuis
	Alopecurus	*A. pratensis*
	Aristida	*A. purpurea*
		A. longiseta
	Calamagrostis	*C. canadensis*
	Muhlenbergia	*M. arenicola*
	M. porteri	
	Oryzopsis	*O. hymenoides*
	Sporobolus	*S. airoides*
		S. cryptandrus
		S. giganteus
	Stipa	*S. comata*
		S. coronata
		S. lettermanii
		S. pulchra
		S. speciosa

TRIBE	GENERA	SPECIES
Agrostidae	*Stipa*	*S. thurberiana*
		S. viridula
Andropogoneae	*Andropogon*	*A. gerardii*
		A. hallii
Aveneae	*Aira*	*A. caryophyllea*
	Arrhenatherum	*A. elatius*
	Avena	*A. fatua*
	Danthonia	*D. parryi*
		D. spicata
	Trisetum	*T. spicatum*
Chlorideae	*Bouteloua*	*B. eriopoda*
		B. gracilis
		B. curtipendula
	Buchloe	*B. dactyloides*
	Chloris	*C. gayana*
	Cynodon	*C. dactylon*
	Eleusine	*E. indica*
	Phleum	*P. pratense*
Festuceae	*Bromus*	*B. catharticus*
		B. grandis
		B. japonicus
		B. laevipes
		B. mollis
		B. tectorum
	Dactyllis	*D. glomeratea*
	Distichlis	*D. spicata*
	Eragrostis	*E. curvula*
		E. lehmaniana
		E. trichodes
	Festuca	*F. idahoensis*
		F. ovina
	Holcus	*H. lanatus*
	Melica	
	Phragmites	*P. communis*
	Poa	*P. bulbosa*
		P. secunda
	Puccinellia	*P. lemmonii*
	Vulpia	*V. microstachys*
		V. octoflora
Hordeae	*Aegilops*	*A. kotschyi*
		A. longissima
	Agropyron	*A. smithii*
		A. spicatum
	Elymus	*E. canadensis*
		E. cinereus
		E. glaucus
		E. junceus
		E. triticoides
	Lolium	*L. multiflorum*
	Sitanion	*S. hystrix*

TRIBE	GENERA	SPECIES
Hordeae	*Taeniatherum*	*T. asperum*
Oryzeae	*Oryza*	*O. sativa*
Paniceae	*Cenchrus*	*C. ciliaris*
	Echinochloa	*E. crusgalli*
	Panicium	*P. obtusum*
		P. dichotomiflorum
	Paspalum	*P. dilatatum*
		P. urvillei
	Pennisetum	*P. americanum*
		P. purpureum
	Setaria	*S. lutescens*
		S. macrostachya
Phalaridae	*Anthoxanthum*	*A. odoratum*
	Phalaris	*P. arundinacea*
Zoysieae	*Hilaria*	*H. belangeri*
		H. jamesii
		H. mutica

Index

drying seeds 48
Dyer's Woad, seed germination 162

Echinochloa crusgalli, seed germination 195
Echinops, seed germination 157
Eichhornia, seed germination 174
Elaegnus, seed germination 87
E. angustifolia 87
E. commutata 87
Elderberry, seed germination 132
Elecampane, seed germination 158
Eleocharis coloradoensis, seed germination 164
Eleusine indica, seed germination 194
Elm, seed collection 21
Elm, seed germination 96
Elymus, seed germination 190
E. cinereus 190
E. glaucus 190
E. junceus 190
E. triticoides 190
embryo 4
embryo excision 70
embryo maturity 58
Encelia virginensis, seed germination 119
endangered species 67
endosperm 4
enhancement of germination 59
environment for seed germination 55
enzyme activation 6
Ephedra, seed germination 119
E. nevadensis 119
E. viridis 119
Epigaea repens, seed germination 119
epigeal germination 5
equipment for seed germination test 55
Eragrostis, seed germination 188
E. curvula 188
E. lehmaniana 188
E. trichodes 188
Erechtites hieracifolia, seed germination 157
Eremocarpus setigerus, seed germination 164
Erigeron, seed germination 157
Eriogonum, seed germination 120
E. arborescens 174
E. cernuum 174
E. crocatum 174
E. fasciculatum 174
E. giganteum 174
E. latifolium 174
Erodium, seed germination 165

Erysimum allionii, seed germination 162
Eschscholzia californica, seed germination 172
ethylene for seed germination 63
Etiophyllum, seed germination 157
E. confertiflorum 157
E. nevinii 157
Eucalyptus, seed germination 87
Euonymus occidentalis, seed germination 120
Eupatorium perfoliatum, seed germination 157
Euphorbia, seed germination 164
E. heterophylla 164
E. marginata 164
Euphorbiaceae, seed germination 164
Eurotia, seed germination 113
Evening Primrose Family, seed germination 171
exploding seed capsules 20

Fagus, seed germination 88
Fairyduster, seed germination 110
Fall Dandelion, seed germination 159
Fallugia paradoxa, seed germination 120
False
Foxglove 178
Hellobore 169
Indigo 108
Fernbush, seed germination 115
Fescue, seed germination 188
Festuca, seed germination 188
F. idahoensis 188
F. ovina 188
Festuceae tribe, seed germination 187
Figwort Family, seed germination 177
Filaree, seed germination 165
Finger Grass, seed germination 193
Fir, seed collection 22
Five-Stamen Tamarisk, seed germination 134
Flax, seed germination 170
Flax Family, seed germination 170
fleshy fruits 32
Flodman Thistle, seed germination 156
Flordia Pusley, seed germination 177
Flowering
Dogwood 86
Flax 170
forage harvesters for seed collection 19
Fouquiera splendens, seed germination 120
Four-wing saltbush, seed collection 21
Foxglove, seed germination 178
Foxtail, seed germination 191

229

Sorbus, seed germination 95
South Dakota Seed Blower 40
Sparganiaceae, seed germination 179
Sparganium, seed germination 180
specific gravity separators 40
Speedwell, seed germination 179
Sphaeralcea, seed germination 101
S. ambigua 171
S. grossulariaefolia 171
Spicebush, seed germination 125
Spike Rush, seed germination 164
Spiny Hopsage, seed germination 122
Spiraea betulifolia var. lucida, seed ger-
    mination 133
Sporobolus, seed germination 192
S. airoides 192
S. cryptandrus 192
S. giganteus 192
Spotted St. John's Wort, seed germina-
    tion 166
Spruce, cone collection 22
Spruce, seed germination 77
Spruge, seed germination 164
Square-stemmed Monkey Flower,
    seed germination 178
Squaw Apple, seed germination 127
Squirreltail, seed germination 190
St. John's Wort Family, seed germina-
    tion 165
standard germination test 69
Stanleya, seed germination 163
S. elata 163
S. pinnata 163
Staphylea bolanderi, seed germination
    133
Star Thistle, seed germination 156
Sterculiaceae, seed germination 180
Stiffbrush Poppy, seed germination
    119
Stipa, seed germination 192
S. comata 192
S. coronata 193
S. lettermanii 193
S. pulchra 193
S. speciosa 192
S. thurberiana 18, 192
S. viridula 192
Stipa thurberiana, seed collection 18
Stonecrop, seed germination 161
stratification 60
methods 61
stripping grass seeds 18
Styrax officinalis, seed germination 133
Suaeda, seed germination 133
substrate for germination test 56

suction harvesters 19
Sugar Maple, seed germination 83
Sugarstick 175
sulfuric acid 59
Sumac, seed germination 130
Summer Holly, seed germination 117
Sunflower Family, seed germination
    154
Sweet
Clover 165
Desert Willow 115
Gum 90
Shrub 110
William 154
Sycamore, seed germination 92
Symphoricarpos, seed germination 133
Syringa, seed germination 134

Taeniatherum asperum, seed germina-
    tion 190
Tagetes, seed germination 160
Tall Oat Grass, seed germination 191
Tallowtree, seed germination 98
Tamarix pentandra, seed germination
    134
Tanoak, seed germination 90
Tansy Mustard, seed germination 102
Taraxacum, seed germination 160
Tarweed, seed germination 158
Taxodium distichum, seed germination
    79
Taxus, seed germination 80
Telegraph Weed, seed germination 158
temperature,
seed storage 48
widely fluctuating 61
testa 5
Tetradymia, seed germination 134
tetrazolium test 70
Texas
Blue Bonnet 168
Persimmon 87
Thalictrum, seed germination 176
T. dioicum 176
T. polycarpum 176
Thistle, seed germination 156
threshing 30
Thrift, seed germination 173
Thuja, seed germination 80
Thurber's Needle Grass 18
Tickseed, seed germination 156
Tilia americana, seed germination 96
timing seed collection 13
Timothy, seed germination 194
Toad Flax, seed germination 178